THE GOD SPECIES

By the same author

High Tide: News from a Warming World
Six Degrees: Our Future on a Hotter Planet

MARK LYNAS

The God Species

How the Planet Can Survive the Age of Humans

FOURTH ESTATE • *London*

First published in Great Britain in 2011 by
Fourth Estate
An imprint of HarperCollins*Publishers*
77–85 Fulham Palace Road
London W6 8JB
www.4thestate.co.uk

2

A catalogue record for this book is
available from the British Library

ISBN 978-0-00-731342-6

Typeset in Minion by G&M Designs Limited,
Raunds, Northamptonshire
Printed and bound in Great Britain by
Clays Ltd, St Ives plc

For my family and other animals

CONTENTS

Acknowledgements 1
Introduction 3

1 The Ascent of Man 15
2 The Biodiversity Boundary 30
3 The Climate Change Boundary 52
4 The Nitrogen Boundary 85
5 The Land Use Boundary 110
6 The Freshwater Boundary 139
7 The Toxics Boundary 157
8 The Aerosols Boundary 183
9 The Ocean Acidification Boundary 198
10 The Ozone Layer Boundary 217
11 Managing the Planet 229

Notes 245
Index 275

ACKNOWLEDGEMENTS

This book is an unusual beast, because it is conceptually and scientifically based on the work of other people. It is traditional in acknowledgements for writers to gush that 'without so-and-so this book would never have been written', but in the case of Johan Rockström and his planetary boundaries co-authors this is literally true. The basic concept is entirely theirs, as are the scientific definitions and quantifications attached to each one. However, the evidence supporting the boundaries presented in this book is largely the fruit of my own research, and the social and economic implications I suggest that the boundaries may carry are my assertions alone. I have no idea whether Johan's co-authors support nuclear power, carbon offsetting or a variety of other controversial proposals that I explore or endorse in this book. They have laid the foundations, however, for a new and better way of looking at our planet, and for that we are all deeply in their debt.

I am particularly grateful to planetary boundaries co-author Professor Diana Liverman, a long-time friend of my family, who invited me to the first scientific workshop on the boundaries concept at the Tällberg Forum in Sweden. That kind gesture set me on this road, and Diana has also been consistently supportive throughout the long process of writing and publishing. As a Visiting Research Associate at Oxford University's Environmental Change Institute (which Diana recently headed) I have also benefited enormously from all the research facilities, expert lecturers and contacts of a world-class university. Most important of all was continuing access to the Radcliffe Science Library, both onsite and remotely: I am deeply grateful to all the helpful staff at the RSL for their assistance and the computing staff who helped keep my VPN connection working through the hard times.

As always, I am indebted to my loyal and very companiable agent Antony Harwood, whose continuing input has been invaluable. The long, multi-year process of writing a book inevitably has its ups and downs, and I could not ask for a better agent or friend. At my publisher Fourth Estate, my editor Robin Harvie – himself an accomplished author – played the invaluable role of helping me see that this book was not quite finished when I thought it was, and has improved the end result immeasurably in the process. Copy-editor Steve Cox's succinct additions and deletions have benefited the manuscript hugely – I can only hope that one day I learn how to get 'that' and 'which' the right way round.

I don't dare embark on a long list of friends and colleagues who have helped in case I leave someone important out. You know who you all are. Thank you. My wife Maria deserves a special mention though, as do my children Tom and Rosa. It is them – and our non-human friends of all kingdoms and phyla – to whom this book is dedicated.

INTRODUCTION

Then Man said: 'Let there be life.' And there was life.

Thunderbolts do not come much more momentous than this: in May 2010, for only the second time in 3.7 billion years, a life-form was created on planet Earth with no biological parent. Out of a collection of inanimate chemicals an animate being was forged. This transformation from non-living to living took place not in some primordial soup, still less the biblical Garden of Eden, but in a Californian laboratory. And the Divine Creator was not recognisably Godlike, despite the beard and gentle countenance. He was J. Craig Venter, a world-renowned biologist, highly successful entrepreneur and one of the first sequencers of the human genome. At the ensuing press conference, this creator and his colleagues announced to the world that they had made a self-replicating life-form out of the memory of a computer. A bacterial genome had been sequenced, digitised, modified, printed out and booted up inside an empty cell to create the first human-made organism. As proof, the scientists wielded photographs of the microscopic '*Mycoplasma mycoides* JCVI-syn1.0' cells, busily obeying the original divine command to be fruitful and multiply in one of the J. Craig Venter Center's many Petri dishes. The new discipline of synthetic biology had come of age.

Forget all your fears about genetic engineering; synthetic biology makes GE look as quaint and old-fashioned as a horse and cart at a Formula One rally. Old-style biotech was about mixing and rearranging small numbers of existing natural genes from different species and hoping that the right thing happened. Synthetic biology is an order of magnitude more powerful, for it gives humanity the potential to design and create life from scratch. Venter and his team didn't quite achieve that: their synthetic genome, after being stitched together with the help of some well-trained yeast, was transplanted into the empty cell of a closely related bacterium that was arguably already 'alive', at least in form if not in function. But the structure the new cells took was that prescribed by the scientists, featuring specially-designed DNA 'watermarks' that included three quotes, the names of the researchers on the project, and an email address for anyone clever enough to successfully decode and sequence the new genome.

The next steps for Venter's team – and other competitors rushing to pioneer novel methods in the same field – point the way towards a new technology of awesome power and potential. Once the function of every gene is understood, scientists can begin to build truly new organisms from scratch with different useful purposes in mind. Microbial life-forms could be designed to create biofuels or new vaccines, to bio-remediate polluted sites or to clean water. In the hands of a modern-day Bond villain, they might also be used to forge virulent new superbugs that could wipe out most of the world's population. But the technology per se is ethically inert; it is just a tool. The purpose of a machine depends upon whose hands are wielding its power. Synthetic biology reduces the cell to a machine, whose components – once properly understood – can be assembled like blocks of Lego. Why build a robot out of perishable steel and plastic when you can build a bio-bot that feeds itself, carries out its prescribed task, heals any injuries, and creates near-identical copies of itself with no outside intervention?

The Book of Genesis is full of instances of Man being punished for his attempts to become like God. After the woman and the serpent combine forces to taste the forbidden fruit from one tree, in Genesis 3:22 the Lord complains: 'See, the man has become like one of us, knowing good and evil; and now, he might reach out his hand and take also from the tree of life, and eat, and live for ever'. Man is banished from Eden to deny him this power of immortality, but Genesis 11:3 once again finds humanity trespassing on the power of the divine, this time with a great tower aimed at reaching Heaven. God's solution to the Tower of Babel was a smart one, achieved by dividing humans into mutually uncomprehending linguistic groups. Today, with the worldwide language of science, that problem has finally been overcome. Venter and his team have seemingly proved that all life is reducible to chemistry – there is nothing more to it than that. No essential life-force, no soul, no afterlife.

With the primacy of science, there seems to be less and less room for the divine. God's power is now increasingly being exercised by us. We are the creators of life, but we are also its destroyers. On a planetary scale, humans now assert unchallenged dominion over all living things. Our collective power already threatens or overwhelms most of the major forces of nature, from the water cycle to the circulation of major elements like nitrogen and carbon through the entire Earth system. Our pollutants have subtly changed the colour of the sky, whilst our release of half a trillion tonnes of carbon as the greenhouse gas CO_2 into the air is heating up the atmosphere, land and oceans. We have levelled forests, ploughed up the great grasslands and transformed the continents to serve our demands from sea to shining sea. Our detritus gets everywhere, from the highest mountains to the deepest oceans: abandoned plastic bags drift ghostlike in the unfathomable depths, even kilometres beneath the floating Arctic ice cap. Wherever you look, this truth is there to behold: pristine nature – Creation – has disappeared for ever.

There is a name for this new geological era. The Holocene – the 10,000-year, climatically equable post-ice age era during which human civilisation evolved and flourished – has slipped into history, to make way for the Anthropocene. For the first time since life began, a single animal is utterly dominant: the ape species *Homo sapiens*. Evolution has equipped us with huge brains, stunning adaptability and brilliantly successful technical prowess. In less than half a million years we have gone from prodding anthills with sticks to constructing a worldwide digital communications network. Who can beat that? Like Venter's bacteria, we have been extremely fruitful and multiplied prodigiously: humans are now more numerous than any large land animal ever to walk the Earth, and the combined weight of our fleshy biomass outstrips that of most other larger animals put together, with the single exception of our own livestock. The productive capacity of a major part of the planet's terrestrial surface is now dedicated to satisfying our demands for food, fuel and fibre, whilst the oceans are trawled round the clock for the fishy fats and proteins our brains and bodies demand. In sum, somewhere between a quarter and a third of the entire planetary 'net primary productivity' (everything produced by plants using the power of the sun) is today devoted to sustaining this one species – us.

With close to 7 billion specimens of *Homo sapiens* currently in existence, mostly enjoying rising (though highly variable) levels of wealth and material consumption, human beings have so far been an evolutionary success story unprecedented in the entire history of planet Earth. But there is a dark side to this momentous achievement. For the biosphere as a whole the Age of Humans has been a catastrophe. Our domestication of the planet's surface to provide crops and animals for ourselves has displaced all competing species to the margins. The Earth is now in the throes of its sixth mass extinction, the worst since the ecological calamity that wiped out the dinosaurs 65 million years ago. Evolution is about competition – and we have outcompeted them all. No other species can control our numbers and return balance to the system (though extremely virulent microbes are likely to come closest). Whenever we have appeared on the verge of shortages, either in food production or fuel for our ever-rising energy demands, we have saved ourselves through brainpower and the judicious application of technology. The worst plague, flood or world war – which may singly or combined cause horrifying loss of life – is just a blip in this relentless upward trend.

But most amazing of all perhaps is how blissfully unaware of this colossal transformation we remain. We are phenomenally, stupendously, ignorant. As if God were blind, deaf and dumb, we blunder on without any apparent understanding of either our power or our potential. Even most Greens – ever hopeful that vanished wild nature can one day be restored – still recoil from the real truth about our role. Climate-change deniers are successful not just because of the moneyed vested interests they serve, but because they tap into a powerful cultural undercurrent that insists we are small and the planet is big, ergo nothing we do – not even in our collective billions

– can have a planet-scale impact. The world's major religions, founded as they were in an earlier, more innocent age, share this insistence, as if the Book of Genesis could still be anything more than a historical metaphor in an era of Earth science and biochemistry. Our culture and politics languishes decades behind our science.

To most people my contention that humans are now running the show smacks of hubris. Consequently everyone loves a good disaster, because it makes us feel small. After the 2004 Asian tsunami there were honest discussions over the benevolence or otherwise of God. Those in the path of hurricanes often speak about the anger of Mother Nature. When the Icelandic volcano Eyjafjallajokull erupted in April 2010, news reports reminded us of 'nature's awesome power over humans', as if a few grounded aircraft in Europe had humbled us helpless clumsy apes. The Japanese earthquake and resulting tsunami disaster in March 2011 showed nature's force at its most powerful and destructive, but many lives were saved because of warning systems and strict building codes. We may not be able to stop earthquakes, but the idea of perennial human victimhood is now somewhat out of date. I suspect there is a reason why most of us cannot bear to let go of it, however, for admitting that we hold the levers of power over the Earth's major cycles would mean having to take conscious decisions about how the planet should be managed. This is an idea so difficult to contemplate that most people simply prefer denial, relieving themselves of any inconvenient burden of responsibility. What you don't know can't hurt you, right?

This see-no-evil approach is particularly convenient for politically motivated climate-change deniers. Take Newt Gingrich, the US Republican firebrand who almost single-handedly destroyed the Clinton presidency and is now taking aim at Obama too. He told the American environment website Grist.org in June 2010: 'It's an act of egotism for humans to think we're a primary source of climate change. Look at what happened recently with the Icelandic volcano. The natural systems are so much bigger than manmade systems.'[1] QED, as I think they say.[2]

Gingrich and his ilk may be an extreme case, but this degree of ignorance and denial cannot go on for much longer. Instead, I suggest that since nature can no longer tame us, then we must tame ourselves. Recognising that we are now in charge – whether for good or ill – we need to take conscious and collective decisions about how far we interfere with the planet's natural cycles and how we manage our global-scale impacts. This is not for aesthetic reasons, or because I mourn the loss of the natural age. It is too late for that now, and – as my uncle always says – one must move with the times. Instead, the overwhelming weight of scientific evidence suggests that we are fast approaching the point where our interference in the planet's great bio-geochemical cycles is threatening to endanger the Earth system itself, and hence our own survival as a species. To avert this increasing danger, we must begin to take responsibility for our actions at a planetary scale. Nature no longer runs the Earth. We do. It is our choice what happens from here.

This book aims to demonstrate how our new task of consciously managing the planet, by far the most important effort ever undertaken by humankind, can be tackled. The idea for it came to me in a moment of revelation two years ago in Sweden, during a conference in the pretty lakeside village of Tällberg. I was invited to join a group of scientists meeting in closed session to discuss the concept of 'planetary boundaries', a term coined by the Swedish director of the Stockholm Resilience Centre, Professor Johan Rockström. The scientists – all world experts in their fields – were trying to nail down which parts of the Earth system were being most affected by humans, and what the implied limits might be to human activities in these areas. Some, like climate change and biodiversity loss, were familiar and obvious contenders for top-level concern. Others, like ocean acidification and the accumulation of environmental toxins, were newer and less well-understood additions to the stable.

During hours of debate, and with much scribbling of numbers and spider diagrams on flip-chart paper, humanity's innumerable list of ecological challenges was reduced to just nine. I left the room late that afternoon certain that something radical had just happened, but not quite sure what it was. It wasn't until later in the evening – in the shower of all places – that I understood in a flash just how important the planetary boundaries concept could be. I realised that scientists studying the Earth system were now in a position to define what mattered at a planetary level, and that this knowledge could and should be the organising basis for a new kind of environmental movement – one that left behind some of the outdated concerns of the past to focus instead on protecting the planet in the ways that really counted. Of course all knowledge is tentative, but here was something very tangible: for the first time, world experts were not just listing our problems, but putting numbers on how we should approach and solve them. I tracked down Johan Rockström and we shared a beer in the hotel lobby. He was encouraging, and we agreed that my job as a writer and as an environmentalist should be to do what the scientists could not: get this scientific knowledge out into the mainstream and demand that people – campaigners, governments, everyone – act on it. Hence this book.

The planetary boundaries concept of course builds on past work conducted by experts in many different fields, from geochemistry to marine biology. But its global approach is actually very new and potentially quite revolutionary. Unlike, say, the 1972 *Limits to Growth* report produced by the Club of Rome, the planetary boundaries concept does not necessarily imply any limit to human economic growth or productivity. Instead, it seeks to identify a safe space in the planetary system within which humans can operate and flourish indefinitely in whatever way they choose. Certainly this will require limiting our disturbance to key Earth-system processes – from the carbon cycle to the circulation of fresh water – but in my view this need constrain neither humanity's potential nor its ambition. Nor does it necessarily mean ditching capitalism, the profit principle, or the market, as many of today's campaigners demand. Above all, this is no time for pessimism: we have some very powerful tools available to allow us to live more gently on this planet, if only we choose to use them.

In this book I take the planetary boundaries concept further into the social, economic and political realms than the original experts were able to. Although some of the planetary boundaries expert group have generously helped to check my facts and figures, I do not expect them to agree with all my suggestions or arguments regarding the implications of meeting the boundaries. There are substantial caveats and uncertainties, as always, and disagreement can be expected between other experts about whether a 'planetary boundary' is truly relevant, and if so, what its limit should be – not to mention how we should meet it. This is first-draft work, Planetary Boundaries 1.0 if you will; there cannot fail to be teething problems. Even so, factual statements in this book are based wherever possible on the peer-reviewed scientific literature – the gold standard for current knowledge. References are at the back, and I urge all readers to make good use of them.

Many will find my analysis and conclusions rather unsettling – not least my colleagues in the Green movement, many of whose current preoccupations are shown to be ecologically wrong. Until now, environmentalism has been mostly about reducing our interference with nature. Central to the standard Green creed is the idea that playing God is dangerous. Hence the reflexive opposition to new technologies from splitting the atom to cloning cattle. My thesis is the reverse: playing God (in the sense of being intelligent designers) at a planetary level is essential if creation is not to be irreparably damaged or even destroyed by humans unwittingly deploying our new-found powers in disastrous ways. At this late stage, false humility is a more urgent danger than hubris. The truth of the Anthropocene is that the Earth is far out of balance, and we must help it regain the stability it needs to function as a self-regulating, highly dynamic and complex system. It cannot do so alone.

This means jettisoning some fairly sacred cows. Nuclear power is, as many Greens are belatedly realising, environmentally almost completely benign. (The Fukushima disaster in Japan did nothing to change this sanguine assessment, and perhaps more than anything reconfirmed it: more on that later.) Properly deployed, nuclear fission is one of the strongest weapons in our armoury against global warming, and by rejecting it in the past campaigners have unwittingly helped release tens of billions of tonnes of carbon dioxide into the atmosphere as planned nuclear plants were replaced by coal from the mid-1970s onwards. Anyone who still marches against nuclear today, as many thousands of people did in Germany following the Fukushima accident, is in my view just as bad for the climate as textbook eco-villains like the big oil companies. (Germany's over-hasty switch-off of seven of its nuclear power plants after the Japanese tsunami will have led to an additional 8 million tonnes of carbon dioxide in just three months.[3]) The same goes for genetic engineering. The genetic manipulation of plants is a powerful technology that can help humanity limit its environmental impact and feed itself better in the process. I personally campaigned against it in the past, and now realise that this was a well-intentioned but ignorant mistake. The potential of synthetic biology I can only begin to guess at today in early 2011. But the lesson is clear: we cannot afford to foreclose powerful technological options like nuclear, synthetic biology and GE because of Luddite prejudice and ideological inertia.

Indeed, if we apply the metric of the planetary boundaries to the campaigns being run by the big environmental groups, we find that many of them are irrelevant or even counterproductive. Carbon offsetting is a useful short-term palliative that the Green movement has discredited without good reason, harming both the climate and the interests of poor people in the process. Some Green groups have also made it very difficult to use the climate-change negotiations as a way to save the world's forests by insisting that rainforest protection should not be eligible for carbon credits. In addition, environmental and development NGOs in general have been much too easy on rapidly emerging big carbon emitters like China and India, whose governments need to be pressed or assisted to eschew coal in favour of cleaner alternatives. Blaming the rich countries alone for climate change may tick all the right ideological boxes, but it is far from being the full story.

Most Greens also emphatically object to geoengineering – the idea that we could consciously alter the atmosphere to counteract climate change, for example by spraying sulphates high in the stratosphere to act as a sunscreen. But the objectors seem to forget that we are already carrying out massive geoengineering every day, as a hundred million people step into their cars, a billion farmers dig their ploughs into the soil, and 10 million fishermen cast their nets. The difference seems to come down to one of intent: is unwitting and bad planetary geoengineering really better than witting and good planetary geoengineering? I am not so sure. At the very least a reflexive rejectionist position risks repeating the mistakes of the anti-genetic engineering campaign, where opposing a technology a priori meant that lots of potential benefits were stopped or delayed for no good cause. Being against something can have just as big an opportunity cost as being for it.

Certainly deciding on something as epochal as intentional climatic geoengineering would involve us in some truly awesome collective decisions, which we have only just begun to evolve the international governance structures to manage. But if we want the Anthropocene to resemble the Holocene rather than the Eocene (roughly 55–35 million years ago, which was several degrees hotter and had neither ice caps nor humans) we will need to act fast. On climate change, meeting the proposed planetary boundary means being carbon-neutral worldwide by mid-century, and carbon-negative thereafter. The former will not be possible in my view without nuclear new-build on a large scale, and the latter will need the deployment of air-capture technologies to reduce the concentration of ambient CO_2. On biodiversity loss, we need to rapidly scale up 'payments for ecosystem services', schemes that use private and public-sector approaches to make planetary ecological capital assets like rainforests and coral reefs worth more alive than dead. To meet the other boundaries we will need to deploy genetically engineered nitrogen- and water-efficient plants, remove unnecessary dams from rivers, eliminate the spread of environmental toxins like dioxins and PCBs, and get much better at making and respecting international treaties. We can learn a great deal from the success of ozone-layer protection, which remains a shining example of how to do it right.

Most importantly, environmentalists need to remind themselves that humans are not all bad. We evolved within this living biosphere, and we have as much right to be here as any other species. Through our intelligence, Mother Earth has seen herself whole and entire for the first time from space[4]. Thanks to us she can even hope to protect herself from extraterrestrial damage: we now operate a programme to track large meteorites like the one that destroyed a significant portion of the biosphere at the end of the Age of Dinosaurs. The Age of Humans does not have to be an era of hardship and misery for other species; we can nurture and protect as well as dominate and conquer. But in any case, the first responsibility of a conquering army is always to govern.

CHAPTER ONE

The Ascent of Man

Three large rocky planets orbit the star at the centre of our solar system: Venus, Earth and Mars. Two of them are dead: the former too hot, the latter too cold. The other is just right, and as a result has evolved into something unique within the known universe: it has come alive. As Craig Venter and his team of synthetic biologists have shown, there is nothing chemically special about life: the same elements that make up our living biosphere exist in abundance on countless other planets, our nearest neighbours included. But on Earth, these common elements – carbon, hydrogen, nitrogen, oxygen and many more – have arranged themselves into uncommon patterns. In the right conditions they can move, grow, eat and reproduce. Through natural selection, they are constantly changing, and all are involved in a delicate dance of physics, chemistry and biology that somehow keeps Earth in its Goldilocks state, allowing life in general to survive and flourish, just as it has done for billions of years.

Why the Earth has become – and has remained – a habitable planet is one of the most extraordinary stories in science. Whilst Venus fried and Mars froze, Earth somehow survived enormous swings in temperature, rebounding back into balance whatever the initial cause of the perturbation. Venus suffered a runaway greenhouse effect: its oceans boiled away and most of its carbon ended up in the planet's atmosphere as a suffocatingly heavy blanket of carbon dioxide. Mars, on the other hand, took a different trajectory. It began life warm and wet, with abundant liquid water. Yet something went wrong: its carbon dioxide ended up trapped for ever in carbonate rocks, condemning the planet to an icy future from which there could be no return.[1] The water channels and alluvial fans that cover the planet's surface are now freeze-dried and barren, and will remain so until the end of time.

Part of the Earth's good fortune obviously lies in its location: it is the right distance from the sun to remain temperate and equable. But the distribution of Earthly chemicals is equally critical: our greenhouse effect is strong enough to raise the planet's temperature by more than 30 degrees from what it would otherwise be, from −18°C to about 15°C today on average – perfect for abundant life – whilst keeping enough carbon locked up underground to avoid a Venusian-style runaway greenhouse. Ideologically motivated climate-change deniers may rant and obfuscate, but geology (not to mention physics) leaves no room for doubt: greenhouse gases, principally carbon dioxide (with water vapour as a reinforcing feedback), are unquestionably a planet's main thermostat, determining the energy balance of the whole planetary system.

This astounding 4-billion-year track record of self-regulating success makes the Earth unique certainly in the solar system and possibly the entire universe. The only plausible explanation is that self-regulation is somehow an emergent property of the system; negative feedbacks overwhelm positive ones and tend to push the Earth towards stability and balance. This concept is a central plank of systems theory, and seems to apply universally to successful complex systems from the internet to ant colonies. These systems are characterised by near-infinite complexity: all their nodes of interconnectedness cannot possibly be identified, quantified or centrally planned, yet their product as a whole tends towards balance and self-correction. The Earth that encompasses them is the most complex and bewilderingly successful system of the lot.

One of the pioneers in understanding the critical regulatory role of life within the Earth system was the brilliant scientist and inventor James Lovelock. Lovelock's original Gaia theory – that living organisms somehow contrive to maintain the Earth in the right conditions for life – was a stunning insight. But his idea of the Earth as being alive, perhaps as a kind of super-organism, only holds good as a metaphor. Self-regulation comes about not for the benefit of any component of the system – living or non-living – but by dint of the overall system's long-term survival and innate adaptability.

An important characteristic of the Earth system is that its main elements move around rather than all ending up in one place. Water, for instance, cycles through rivers, oceans, ice caps, the atmosphere and us. An H_2O molecule falling in a snowstorm on the rocky peak of Mount Kenya may have been exhaled in the dying gasps of Queen Elizabeth I: water, driven by energy, is always circulating. Nitrogen, oxygen, phosphorus, sodium, iron, calcium, sulphur and other elements are also perpetually on the move. Carbon is perhaps the most important cycle of all, because of the thermostatic role played by its molecular state; particularly in its gaseous form as CO_2, but also in combination with other elements, such as with hydrogen as CH_4 (methane). It was the failure of the carbon cycle that doomed Venus and Mars, yet here on Earth various feedbacks have kept the system in relative balance for billions of years – even altering the strength of the greenhouse effect to offset the sun's increasing output of radiation over geological time.

Over million-year timescales, the carbon cycle balances out between the weathering of rocks on land, which draws carbon dioxide out of the air, and its emission from volcanoes. Carbon is deposited in the oceans and then recycled through plate tectonics, as oceanic plates subduct under continental ones, providing more fuel for CO_2-emitting volcanoes. The process is self-correcting: if volcanoes emit too much carbon dioxide, the Earth's atmosphere heats up, increasing weathering rates and drawing down CO_2. If carbon dioxide levels fall low enough for weathering to cease – as perhaps was the case during the early 'snowball Earth' episodes, when global-scale ice caps put a stop to the weathering of rocks – volcanic emissions continue uninterrupted, allowing CO_2 to build up until a stronger greenhouse effect melts the ice and allows balance to be restored. The system is stable but not in stasis: the geological record shows tremendous swings in temperature and carbon dioxide concentrations over the ages, though always within certain boundaries.

Perhaps one of the strongest arguments against the Gaia concept is the fact that even if the planet in general remains habitable, things do sometimes go badly wrong. Over the last half-billion years since complex life began there have been five serious mass extinctions, the worst of them wiping out 95 per cent of species alive at the time. Most appear to have been linked to short-circuits in the carbon cycle, where volcanic super-eruptions led to episodes of extreme global warming that left the oceans acidic and depleted in oxygen, and the land either parched or battered by merciless storms. And yet, over millions of years, new species evolved to fill the niches vacated by extinguished ones, and some kind of balance was restored. Over the last million years, recurrent ice ages demonstrate how regular cycles can lead to dramatic swings in temperature, as orbital changes in the Earth's motion around the sun lead to small differences in temperature, which are then amplified by carbon-cycle and ice-albedo (reflectivity) feedbacks. Our planet may be self-regulating, but it is also extraordinarily dynamic.

GOD SPECIES OR REBEL ORGANISM?

Life is now an important component of most of the planet's major cycles. The majority of carbon is locked up in calcium carbonate (limestone) rocks, laid down in the oceans by corals and plankton. The appearance of photosynthesis was perhaps one of life's most miraculous innovations, allowing microbes – and later, green plants – to use atmospheric carbon dioxide as a source of food. Water is an essential part of the process: in cellular factories called chloroplasts, plants split water into hydrogen and oxygen, combining the hydrogen with carbon from the air to form carbohydrates, and releasing oxygen as a waste product. The process opened up an opportunity for the evolution of animals, that could eat the carbohydrates as a food source and recombine them with oxygen (forming CO_2 and water), thereby generating energy and closing the loop.

Evolution of life is a critical part of the process of planetary self-regulation, because it allows organisms to change to take advantage of new opportunities and learn from failures – evolution is self-correction in action. Just as the build-up of oxygen in the air allowed animal life to appear, so the accumulation of any waste is an opportunity for new species to evolve to take advantage of it. Evolution is very different from mere adaptability, because it allows new life-forms to appear rather than old ones to adapt, leading to much greater transformations. A species may, for example, be able to adapt to a shift in its food supply by moving, but over many millennia an entirely new species may thereby come into being, able to exploit a whole new niche in the ecosystem. Think of polar bears, likely descended from an isolated population of brown bears in an ice age, but which evolved white fur and an ice-based lifestyle to become the pre-eminent hunter of the far north.

All this sounds comforting. The Earth, and life, will always prevail. But the self-regulating system contains a flaw, one that can seriously damage or even destroy it. This flaw is the gap in time between a perturbation and the ensuing correction: instabilities can happen very fast, whilst the correcting process of self-regulation typically takes much longer. The gap between the advent of an oxygen-rich atmosphere and the appearance of animal life was a long one: a good hundred million years if not more. Major volcanic eruptions may release trillions of tonnes of carbon dioxide over just a few thousand years, outstripping the capacity of the Earth system to mop up the additional CO_2 via rock weathering and other processes of sequestration, and leading to extreme global warming events. Mass extinctions happen because changing circumstances outstrip the adaptability of existing species before evolution can work its magic. Over millions of years new species can appear, but only from the diminished gene pool of the survivors – and a return to true pre-extinction levels of biodiversity may take much longer, if it ever takes place at all.

This time-lag effect was cleverly demonstrated in a modelling simulation undertaken by two British researchers, Hywell Williams and Tim Lenton (both at the University of East Anglia; Lenton is a member of the planetary boundaries expert group).[2] In a computer-generated world – entirely populated by evolving micro-organisms living in a closed flask – Williams and Lenton found that the closing of nutrient loops emerged as a robust property of the system nearly every time the model was run. As in the real world, the emergence of self-regulation came about because evolution allowed new species to appear that could use the waste of one species as food for themselves, recycling nutrients and leading to a stable state. Moreover, the more species that evolved, the greater the amount of recycling and the greater the overall biomass the system could support. 'Flask world' had discovered the value of biodiversity.

But this world also had a dark side, for several simulations illustrated that the flaw in self-regulation – the time gap between a disturbance and the evolved correction – might occasionally be fatal. In just a few model runs, an organism appeared that was so spectacularly successful in mopping up nutrients that its numbers exploded and its wastes built up to toxic levels before other organisms were able to evolve a response. Williams and Lenton dubbed these occasional rogue species 'rebel organisms'. They were unusual, but their impact was invariably catastrophic: the explosive initial success of the rebels changed the simulated global environment so suddenly and dramatically that their compatriots were killed, and – with no other life-forms around to recycle their wastes – they were themselves condemned to die too. As the last lonely rebels perished, their whole biosphere went extinct, evolution ceased, self-regulation failed, and life wiped itself out.

Like Lovelock's Gaia, Flask world – and its rebel organisms – might just be a clever idea, more of a metaphor than a true representation of reality. But the parallels with our species are unsettling. We have transformed our environment within just a few centuries in ways that are wiping out other life-forms at a shocking speed – the changes so rapid that evolution has no time to adapt and thereby allow other organisms to survive. Like a rebel organism, our species discovered a colossal new source of energy, which had lain hidden and undisturbed for millions of years, and which no previous life-form had found a use for. It is the sheer rapidity in the rise of the waste from the exploited new energy source of buried carbon – largely in the form of gaseous carbon dioxide – plus the other combined wastes and environmentally transformative impacts that fossil fuels allowed humanity to achieve, that have now begun to overwhelm the self-regulatory capacity of the Earth system. This single element holds the key to a possible future mass extinction.

Flask world is now our world. Consider that our wastes are accumulating so fast in the oceans that no species can consume them; instead, massive dead zones are spreading around the world's coasts, from China to the Gulf of Mexico, where the recent BP oil spill adds to the toll. We have produced novel organic chemicals and synthetic polymers that no microbes have yet learned to digest, and which are poisonous to most organisms – often including ourselves. And we are steadily eating our way through global biodiversity – from fish to frogs – consuming voraciously, and moving on to the next species when one is extinguished. Those species that are not edible we ignore and displace, whilst those that threaten or dare to compete with us we pursue mercilessly and annihilate. Thus is our rebel nature revealed.

There is a paradox however. Even as a putative rebel organism, humanity is a product of Darwinian evolution, like every other naturally generated life-form sharing our planet today. Moreover, we did not evolve the biological capacity to eat coal and drink oil – the energy from these abundant 'nutrients' is combusted outside the body rather than metabolised within it. Why us, then? Our mastery of fire was a product of the adaptability and innovativeness with which evolution had already equipped us long before, and that no other species had heretofore possessed. Humanity's Great Leap Forward was not about evolution, but adaptation – and could therefore move a thousand times faster.

I don't want to oversimplify: the Stone Age did not end in 1764 with James Watt's invention of the steam engine. Clearly great leaps in human behaviour and organisation took place over preceding millennia with the advent of language, trade, agriculture, cities, writing and the myriad other innovations in production and communications that laid the foundations for humanity's industrial emergence. But I would argue that the true Anthropocene probably did begin in the second half of the eighteenth century, for it was then that atmospheric carbon dioxide levels began their inexorable climb upwards, a rise that continues in accelerated form today. This date also marks the beginning of the large-scale production of other atmospheric pollutants and the planet-wide destabilisation of nutrient cycles that also characterise this new anthropogenic geological era.

Take population. When humans invented agriculture, some 10,000 years ago, the global human population was somewhere between 2 and 20 million[3]. There were still more baboons than people on the planet. By the time of the birth of Christ, the globe supported perhaps 300 million of us. By 1500, that population had increased to about 500 million – still a relatively slow growth rate. A global total of 700 million was reached in 1730. Then the boom began. By 1820 we numbered a billion. That total rose to 1.6 billion by 1900, and the growth rate continued to accelerate. By 1950 we were 2.5 billion strong, and by 1990 had doubled again to more than 5 billion. In 2000 the 6 billion mark was passed. At the time of writing, in late March 2011, we number an astonishing 6.88 billion individuals.[4] Through the process of writing this book, another 225 million people were added to the total – just under half the entire world population of 500 years ago, now appearing in just three years.

But this still doesn't answer the puzzle: Why us? And why were buried stores of carbon the 'nutrients' that allowed our species to proliferate so explosively? A satisfactory response requires a brief digression into the evolutionary origins of this remarkable hominid, for it is our past that holds the key to our present and future. This is the story of a species whose biological characteristics combined with an accident of fate to have world-shattering consequences. And it is a story that might shed some light on the central question of this book – whether we are rebel organisms destined to destroy the biosphere, or divine apes sent to manage it intelligently and so save it from ourselves.

Perhaps the environmentalist and futurist Stewart Brand put it best when he wrote these words: 'We are as gods and have to get good at it.'[5] Amen to that.

? typo

THE DESCENT OF MAN

Listening to some environmentalists talk, it is easy to get the feeling that humanity is somehow unnatural, a malign external force acting on the natural biosphere from the outside. They have it wrong. We are as natural as coral reefs or termites; our inherited physiology is entirely the product of selective pressures operating over millions of years within living systems. Our inner ear, for example, was once the jawbone of a reptilian ancestor. Babies in the womb begin life with tails, expressing in the earliest stages of life genes that illustrate our long evolutionary history. Our key biological characteristics – including those that have allowed us to emerge as 'sapient' beings – exist only because they conferred on our ancestors some selective advantage as they ate, fought, played and reproduced over millions of years within the natural biosphere.

The actual origin of life – how animate organisms assembled themselves out of inanimate chemicals without a Dr Venter to supervise affairs – remains a mystery. Perhaps the first self-replicating amino acids were formed in some primordial soup by a charge of lightning or a volcanic eruption. Or maybe, given the right environment and ingredients, life can spontaneously appear. Some suggest that extraterrestrial microbes may have hitched a lift onto the early Earth from passing meteors or comets. Either way, the first microbes appeared about 3.7 billion years ago, evolving into 'eukaryotic' cells – with a proper nucleus, cell walls and the capacity to metabolise energy – a billion and a half years later. These cells were probably made up of a symbiotic union of several bacteria, which is why mitochondria in our body cells today still have their own DNA. (Symbiosis, by the way, is quite as much part of the story of evolution as red-in-tooth-and-claw competition.)

Some of these early microbes, the cyanobacteria, learnt to use photons from the sun to split water and carbon dioxide in photosynthesis. They are probably Earth's most successful organisms, for cyanobacteria are still prolific today. As eukaryotic cells learned to combine to form multicellular organisms, the stage was set for a major proliferation of life – though still only in the oceans – in an event dubbed the 'Cambrian explosion' by palaeontologists. During the Cambrian, from 540 million years ago, recognisable ancestors of many of today's animal groups appeared. These include arthropods (insects, spiders and crustaceans), molluscs (snails, oysters, octopus), and even early vertebrates – the first fish. An evolutionary arms race kicked off, as predators evolved ways to catch, grip and swallow, whilst prey developed speed or armour to reduce their chances of being eaten.

Of all the technical novelties evolution called into existence, from scales to jaws, perhaps the most interesting is the development of sight. The eye may have been the innovation sparking this intense burst of Cambrian competition, for both predators and prey would have had an equally powerful reason to evolve vision. The fossil record demonstrates that sight evolved independently in different groups of animals, though in a remarkably similar way. The octopus, for example, has an eye much like ours, with a lens and a retina behind it, yet our common ancestor was probably some kind of sightless worm. All the higher animals that survived the Cambrian could see.

The oceans now had a fully developed food web, and it may have been to escape the marine killing fields that some of the less well-armoured fish first ventured onto land – already colonised, from about 450 million years ago, by plants and insects. Fins gradually morphed into limbs, though the hybrid water–land transition is still repeated in the life cycles of today's amphibians, hundreds of millions of years later. As some of these early amphibians grew more accustomed to onshore life, they evolved into reptiles, with leathery skins to hold in moisture and eggs with watertight shells that could be laid on dry land rather than in ponds.

We are now up to 300 million years ago in geological time – nearly to the appearance of mammals, for our mammalian line is surprisingly ancient, if rather insignificant for most of its existence. The sail-back reptile *Dimetrodon* displayed many mammal-like features: its sail was probably a way to regulate temperature, perhaps demonstrating an early attempt at warm-bloodedness. Its teeth had differentiated into molars and canines, just as ours still do. Its descendants developed fur, modified – like the feathers of birds – out of reptilian scales, also as a way to control its body temperature. By the late Triassic, true mammals appeared, and were present on Earth throughout the entire age of the dinosaurs, though as very junior partners indeed. For the next 135 million years – during the entire Jurassic and Cretaceous periods – our ancestors stayed in the shadows, living furtive existences as the dinosaurs dominated the planet.

Mammals then were tiny, most no bigger than rats. They could dart out under the cover of darkness, snatching insects and worms as *Tyrannosaurus* slept. But there was an evolutionary tradeoff. Without the luxury of laying masses of eggs, and confined to burrows and crevices, mammals evolved sophisticated ways of nurturing their young: live births and lactation. Their specialised teeth enabled them to chew and grind up food, yielding more energy. In contrast the bulky dinosaurs wolfed their meals down whole. But the most outstanding adaptation of the mammals to their subordinate status was far more important than milk or molars. It was the evolution of intelligence. Contrary to popular myth, dinosaurs had big brains – not because they were smart, rather because they were big animals. But it is not brain size per se that counts for intelligence; more important are the relative proportions of brain and body, and in the diminutive mammals, this relationship was beginning to change. As one evolution textbook puts it: 'The pint-sized mammal was the intellectual giant of its time.'[6]

So why did selective pressures force this shift? Most likely, the shadowy existence of mammals demanded very different skills from those of the daytime excursions of dinosaurs. The mammalian world was one of sound and smell as much as sight, demanding more subtle skills of deduction and reasoning. The smell of a predator, for instance, could mean danger if the killer is soon to return – or safety if it is gone. All would need to be kept in memory for retrieval later. Similarly, to interpret sound on a dark night would require consulting a mental map of some complexity, adding further evolutionary pressure for larger brains. The result was the neocortex, a completely new brain structure found only in mammals. This is our 'grey matter' – vital for all higher functions that we collectively define as 'intelligence', such as sensory perception, spatial reasoning and conscious thought.

The age of mammals dawned, with spectacular suddenness, 65 million years ago. Perhaps aggravated by extensive volcanic eruptions and consequent global warming, a mass extinction tore through the planetary biosphere when a large asteroid ploughed into the sea off modern-day Mexico. Once the dust had settled, the dinosaurs were gone – along with half of life on Earth. Why mammals made it through the bottleneck, no one knows. Perhaps they were better protected from the environmental holocaust thanks to their furry, furtive existences. Either way, the end-Cretaceous extinction cleared the way for the explosive evolution of mammals into all the ecological niches previously occupied by dinosaurs. Some took to the water, losing their four legs and re-evolving the fins they had lost over 300 million years earlier to become dolphins and whales. Others joined birds in the air, the fingers of their 'hands' splaying out to form wings, becoming bats. Still more returned to herbivory, and headed out into the grasslands now spreading through the continents, their bodies growing rapidly in size: these became bison, elephants, horses and other grazing and herding animals.

But our story follows a different group of mammals who struck out in a new direction. They headed off not into the land or out to sea but up the trees. Perhaps to escape predators on the forest floor, or to take advantage of succulent arboreal fruits, the lives of these 'prosimians', who appear in the fossil record about 55 million years ago, demanded a whole new set of skills. The paws of their ratlike ancestors evolved into gripping hands, more suited to a life spent grasping branches. Their requirement for smell declined. But their need for vision increased enormously, and not just any vision: their eyesight had to reveal excellent colour, and, most important, had to be front-of-head and stereoscopic to give depth perception.

The pressure was on for bigger brains. Mental calculations performed whilst speeding through the treetops had to be fast and accurate. Memory was once again useful, aiding decisions as to what types of trees could support what weight, how to grasp certain branches, or when to visit different fruiting bits of the forest. These were still small animals, but as they evolved better agility in the forest, their bodies grew larger. By 35 million years ago, true monkeys had appeared. By 22 million years ago, gibbons had split away from the evolutionary line. Orang-utans followed, at about 16 million years ago, and chimpanzees 6 million years ago. That left the hominids, and we are their only surviving descendants – all other hominid species, of which there have been a dozen at least, were destined to perish.

BIRTH OF THE FIRE-APE

Our lineage may be ancient, but modern *Homo sapiens* has been a very short-lived phenomenon, perhaps illustrating the biological anomaly that we are. Although bipedal hominids were stalking the African plains as long as 3 million years ago, true *Homo sapiens* – the evolutionary descendant of *Australopithecus*, *Homo habilis* and later *Homo erectus* – appeared less than 500,000 years ago, and perhaps as recently as 200,000 years ago.

Mitochondrial DNA passed through the maternal line suggests in fact that we are all descended from a single individual – the so-called Mitochondrial Eve – who lived in Africa 200,000 years ago. Further evidence comes from the remarkable homogeneity of human DNA: despite superficial differences in hair straightness, noses and skin colour, we are far more closely related than might be expected. (A single breeding group of chimpanzees will show more genetic variation than do all humans.[7]) This is strong evidence that modern humans did all descend from the same original group, and our dominance may have begun with a characteristic act of genocide, as the last *Homo neanderthalensis* survivors were ethnically cleansed from Europe and Asia by the new migrants. Since then, no other animal, whether on two legs or four, has challenged the dominance of *Homo sapiens*.

The most striking biological characteristic of the human ancestral line over the last few million years is the extraordinary progress of its brain development. Chimpanzee brains measure about 360 cubic centimetres in volume. Early *Australopithecus* had expanded its brain to about 500 cm^3, whilst *Homo erectus* measured up with a brain size of about 800 to 900 cm^3. Half a million years ago, the brain was expanding at an extraordinary rate of 150 cm^3 every hundred thousand years.[8] Modern humans typically have a brain size of 1,350 cm^3, nearly four times the size of those of our nearest relatives, the chimpanzees.

One human innovation is often neglected in accounts of our evolution – and it may be one of the most important of all, because it allowed us to fuel our process of encephalisation (increased braininess). The brain is a very energy-hungry organ, consuming a quarter of all our energy use, as compared with 10 per cent in other primates and 5 per cent in most mammals.[9] So how were the extra food requirements satisfied? Part of the answer is almost certainly the increasing amounts of animal protein in the human diet – hominid species quickly supplanted leopards as the dominant hunters on the African plains. But just as important was the advent of cooking, which enables food to be transformed into much softer and more calorific forms before being eaten. For over a million years humans have been eating cooked food, giving us a dietary advantage no animal has ever enjoyed before.

Cooking, of course, needs fire. Indeed there is a strong biological case for seeing humans as a co-evolved fire species. Fire made us physically what we are, by allowing us to grow vastly bigger brains through eating cooked food. The human gut is much smaller, and uses far less energy, than the digestive system of comparable animals. We also have weak jaws, small mouths and underdeveloped teeth compared with other primates. That first acquisition of fire acted as a powerful evolutionary driver, enabling humans to become the first truly sentient beings in history.

Fire, however, is a very special tool. Not for nothing is it identified in many human cultures as the preserve of the gods. Bonfires lit at the Celtic festival of Beltane symbolise the return of the sun to warm the Earth after the freezing nights of winter. In Navajo tradition, Coyote – who was a friend of humans – tricked two monsters on 'fire mountain' into letting him light a bundle of sticks tied to his tail, which he then took back to people. Perhaps the best-known fire tale of all is that of Prometheus, the Titan of the ancient Greeks (and son of Gaia, goddess of the Earth), who stole fire from the supreme god Zeus and brought it back to people. For this transgression he was punished by being chained to a rock and having his liver eaten out each day by an eagle.

And rightly so, for fire dramatically changed our relationship with the natural world. Acquiring the power of gods separated humans permanently and irretrievably from all other species. As well as cooked food, it afforded protection against predators and warmth on cold nights, allowing early humans to spread north out of Africa during the depths of the last ice age. Fire may have facilitated the spread of genes for hairlessness, as the need for body insulation diminished. However, once our hair was lost and our guts had shrunk, we were tied to the hearth – we could no longer exist without it.

No human can hope to survive in the wild today without fire, and this dependence marks a major qualitative shift in human relations with the biosphere. Other animals need only food. We are the only animal that has learned to harness an external energy source in a systematic way, through our reliance on fuel. It is this food–fuel relationship that most defines the fire-ape, *Homo pyrophilus*. Moreover, this innovation was perhaps the most important one in unbalancing our relationship with nature, for being armed with fire put the rest of the world at our mercy.

However, our dependence on fuel could also be a weakness. Once the forests were chopped down and the landscape denuded, humans might no longer be able to flourish. The story of the modern era, however, is the story of our transcendence over even this limitation. For modern humans were to discover a new source of fuel that would allow us to expand both our numbers and our dominance dramatically. This new fuel, in the form of underground deposits of fossilised biological carbon, was to be the energy springboard that catapulted our species – and the planet – into an entirely new geological era, the Anthropocene. Using the tool of the gods, we were to become as gods. But unlike Zeus, we still live in ignorance about our true power. And time is running out, for the flames of our human inferno have begun to consume the whole world.

CHAPTER TWO

The Biodiversity Boundary

Our fire-sticks and engines have turned humans into extremely successful predators. We have poisoned, outcompeted or simply eaten so many other species that the Earth is currently in the throes of its most severe mass extinction event for 65 million years, and it is this crisis of biodiversity loss that arguably forms humanity's most urgent and critical environmental challenge. Many of our other impacts on the Earth system are more or less reversible, but extinction is for ever, and a flourishing diversity of life is essential for the biosphere to function successfully during the Anthropocene and beyond. By removing species, we damage ecosystems, collapse food webs and ultimately undermine the planetary life-support system on which our species depends just as much as any other.

The planetary boundaries expert group proposes a biodiversity loss boundary of a maximum of ten species lost to life per million species per year. The current rate of loss is already one or two orders of magnitude greater than this: conservationists estimate that 100 to 1000 species per million are currently wiped out annually. Meeting this boundary target is possible, but to do so will require a massive increase in the global attention and funding given to the issue and to solving it. We must create many more nature reserves, both on land and at sea. We must properly fund conservation, to defeat poachers and protect wildlife from direct threats. Above all we must alter our accounting systems so that living systems – from rainforests to polar tundra – are given the value they deserve as literally priceless assets of natural capital. This means using the power of markets, with most payments for biodiversity protection going to the local people who are always the best custodians of their local environment.

If we are to save what remains of the glorious diversity of life on Earth, we will have to act fast. A quarter of the world's mammals, a third of amphibians, about 13 per cent of birds, a quarter of warm-water corals, and a quarter of freshwater fishes are globally threatened with extinction. The rate of loss is accelerating, despite increasing concern about this brutal devastation of our planet's natural history: whilst 36 mammals improved in terms of how threatened they were between the 2007 and 2008 Red Lists, 150 saw a deterioration, from vulnerable to endangered, from endangered to critically endangered, or critically endangered to extinct.[1]

In 2002 world governments agreed a target 'to achieve by 2010 a significant reduction of the current rate of biodiversity loss at the global, regional and national level as a contribution to poverty alleviation and to the benefit of all life on Earth'. Laudable, of course. So was it met? Not even close. To put numbers on the current crisis, a recent report in *Science* looked at 31 different indicators – things like habitat quality, population trends, extinction risk and so on – and found that virtually all of them were either getting worse or showed no improvement in the last decade.[2] Stalin said that the death of one person was a tragedy, the death of millions a statistic (and he was an expert). So it seems for species too. Each story is a unique tragedy, yet the aggregated numbers somehow fail to convey the magnitude of this loss.

Even where absolute extinction has been avoided, many species have become functionally extinct in the sense that their remaining numbers are so few – or so scattered – that they no longer play any effective part in the ecosystem. The Iberian lynx, for example, is not extinct – not yet – in the wild, but its total population (between 84 and 143 adults; split into two isolated populations in Spain) is so tiny that it can hardly still be considered the apex predator it once was. (The lynx may be completely extinct in neighbouring Portugal: its survival has been inferred only by the discovery of a single dropping, identified by molecular analysis in 2001.[3]) Globally the abundance of vertebrate species fell by nearly a third between 1970 and 2006, according to the 2010 Global Biodiversity Outlook.[4] Forget about extinctions: there are now *a third fewer wild animals* in total on the planet than there were forty years ago. That really is a shocking figure.

Even emblematic species like the tiger have their backs to the wall. Globally, only about 3,500 wild tigers remain – an extraordinary statistic given the charisma and recognition factor of this species, whose form has been emblazoned on everything from cereal packets to petrol stations. Three subspecies, the Bali, Caspian and Javan tigers, are already extinct; the South China tiger has probably joined them, for no one has seen it in the wild for 25 years.[5] According to a late-2010 study, the decline in tiger numbers 'has continued unabated' for the last two decades: only 1,000 breeding females now survive, over less than 7 per cent of their historical range. Several Indian so-called 'tiger reserves' no longer have any tigers in them at all. Yet saving the tiger could cost as little as $82 million per year, according to one estimate – this is all it would take to protect the remaining 42 sites around Asia where viable tiger populations remain.[6] All that is needed is a mechanism to raise the funds and an implementation plan to safeguard the reserves.

Particularly badly hit by our success have been our nearest relatives, the great apes. All are threatened with extinction in the wild. In Asia the orang-utan – once common from South China to the Himalayas – is now reduced to a remnant of between 45,000 and 69,000 individuals, mostly in the sort of lowland forests in Borneo that seem to be particularly irresistible to oil-palm plantation owners. In Africa the famous 'gorillas in the mist' of Virunga National Park in the Congo are down to about 380 individuals, under siege by marauding rebels as well as by poachers and bushmeat hunters. To put humans in our proper context, try entering 'great apes' into a www.iucnredlist.org (a website run by the International Union for the Conservation of Nature, featuring its Red List of endangered species) search. When I tried, the results were as follows:

Gorilla beringei (Eastern Gorilla) – Status: Endangered, Pop. trend: decreasing.
Gorilla gorilla (Lowland Gorilla) – Status: Critically Endangered, Pop. trend: decreasing.

***Homo sapiens* (Human) – Status: Least Concern, Pop. trend: increasing.**
Pan paniscus (Gracile Chimpanzee) – Status: Endangered, Pop. trend: decreasing.
Pan troglodytes (Common Chimpanzee) – Status: Endangered, Pop. trend: decreasing.
Pongo abelii (Sumatran Orang-utan) – Status: Critically Endangered, Pop. trend: decreasing
Pongo pygmaeus (Bornean Orang-utan) – Status: Endangered, Pop. trend: decreasing

As this list shows, we are just apes. But with our newfound global power comes a responsibility for proper global stewardship. This is a new task for humans to take on, certainly at a planetary level. But the time for this shift is long overdue, for a brief review of our history to date shows us in a very singular role: that of serial killers.

THE PLEISTOCENE OVERKILL

Many thousands of years ago a dramatic ecological calamity began to sweep through the fauna that inhabited the Earth's disparate continents. Australia lost most of its large animals first, about 46,000 years ago. North and South America saw a similar extinction wave 13,000 years ago. New Zealand, meanwhile, kept hold of its big-bodied animals until a mere 700 years ago. What happened at each of these points in time? Did the climate perhaps change, leaving large animals stranded? Unlikely: there is no correlation between global climate change and the various extinction pulses. Did a meteor strike or a volcano blow? Again, there is no way to pin all of these different calamities, taking place at very different times, on a single geological event. Indeed, the true nature of this extinction calamity is much more familiar. It came on two legs, for a start. What links these points in time is simple: they mark the moment when humans arrived.

Modern humans have at least dealt out death fairly: we began our existence by killing each other. In what looks like a prehistoric bout of all-too-modern ethnic cleansing, *Homo sapiens* probably drove its closest hominid relatives, *Homo neanderthalensis* and *Homo erectus*, to oblivion. A minority of archaeologists cling to the notion that some interbreeding must have taken place, but genetic studies show this is unlikely.[7] Modern human DNA instead confirms that all of us are descended from the same small initial *Homo sapiens* population that migrated out of Africa 50,000 years ago.[8] The last Neanderthals hung on in remote mountainous parts of France until 38,000 years ago, and in southern Spain until about 30,000 years ago. The very last families died a few thousand years later in Gorham's Cave in what is now Gibraltar, when their final refuge on the extreme southern edge of the continent was overrun.[9] Officially, the direct cause of their ultimate demise is a mystery, but I think we can guess who the culprit was.

There is certainly enough evidence to mark out a crime scene. One Neanderthal skeleton discovered in Iraq bears a peculiar puncture wound on one of its ribs – a mortal injury that is most consistent with a spear thrown by an anatomically modern *Homo sapiens*.[10] In early 2009, the anthropologist Fernando Rozzi reported the discovery of a Neanderthal child's jawbone, found together with anatomically modern human remains at the cave of Les Rois in southwestern France.[11] The bone bore characteristic cut marks, similar to those found on butchered reindeer skulls, suggesting that the tongue had been cut out and eaten. Some loose teeth scattered around also had holes drilled in them, perhaps as parts of a morbid ceremonial necklace. Rozzi drew an unequivocal conclusion: 'Neanderthals met a violent end at our hands, and in some cases we ate them,' he said.[12]

There is even stronger evidence surrounding who killed most of the world's largest animals, for their butchered bones are found stacked up everywhere humans invaded. As palaeontologist Richard Cowen writes in *The History of Life*, 'From Russia to France, [archaeological] sites contain the remains of thousands of horses and hundreds of woolly mammoths.'[13] But the slaughter was far worse in the New World, where native species had no previous experience of this naked and harmless-looking but surprisingly rapacious two-legged predator. The North American death toll included six species of ground sloths, two species of mammoths, all mastodons, a giant bison, seven species of deer, moose and antelope, three species of tapirs, the North

American lion, the dire wolf, the giant anteater, the giant turtle, the giant condor, all ten species of North American horses (then absent until reintroduced by invading sixteenth-century Europeans), two species of sabre-toothed cats, eight species of cattle and goats, the North American cheetah, four species of camels and two species of large bears.

But the biggest wipe-out of all took place in Australia, which saw a near-total extinction of large wild animals. The continent lost some extraordinary creatures: a gigantic horned turtle as big as a car, enormous flightless birds standing more than 2 metres tall and weighing half a tonne, a snake 6 metres long, and a giant predatory lizard that grew up to 7 metres in length and must have been the most fearsome reptilian predator since the dinosaurs. About twenty species of large marsupial disappeared, including a cow-sized wombat and a kangaroo 3 metres high. Quite how and when they died remains controversial: many archaeologists have tried to absolve *Homo sapiens* of the crime, pointing to the lack of kill sites and the low density of human population. But the extinction is roughly coincident with human arrival in the continent, and the pattern – affecting the largest species disproportionately – is exactly the same as everywhere else.

Further damning evidence comes from Tasmania, which retained its giant kangaroos (and various other megafauna) for four thousand more years, until falling sea levels allowed humans to finally invade – whereupon the island's giant kangaroos (amongst six other large-bodied species) promptly died out.[14] Any remaining doubters need only look to New Zealand. When Polynesian people first arrived by boat a mere 700 years ago, they found a unique island ecosystem where – thanks to millions of years of geographical isolation – birds rather than mammals or reptiles had evolved to become the dominant land animals. Giant flightless moas stalked the forests, whilst enormous eagles, the largest ever known, with wingspans of the order of 3 metres, soared above the mountains. Within as little as a century all – along with half of the islands' other terrestrial vertebrates – were dead.[15] This time there can be no dispute as to the cause of death or the identity of the killers, for Maori dwelling sites are surrounded by piles of moa bones – some so extensive that they have since been quarried for fertiliser. No doubt believing that the abundance of their moas would last for ever (another pattern that keeps repeating itself), the Maoris wastefully ate only the upper legs and threw the rest away.[16]

Only one continent's large animals survived relatively unscathed. That continent was Africa, whose megafaunal inhabitants had co-evolved with hominids over millions of years and had therefore acquired a great deal of useful experience about living with *Homo sapiens*. As a result, Africa gives us the best idea of what a pre-human landscape might have looked like, with big animals like elephants browsing the undergrowth and herds of wild horses and cattle stirring up dust clouds across the savannah. Indeed, African ecosystems have been used as a model for proponents of 'rewilding' parts of North America; if cheetahs, elephants and camels can be imported into places like Montana, perhaps they could assume the ecological niches vacated by their extinct relatives, some have suggested.[17] This is a romantic but vain hope, not least because the ancient homeland of these large surviving animals is seriously endangered by today's generations of human beings. Africa is safe no more.

Right across the world, these lost big animals left 'ghost habitats' behind – trees that still bear specialised fruits hoping some long-gone giant will distribute them, or thorny bushes protecting themselves against browsing by extinct large herbivores. In Brazil, more than 100 tree species still produce obsolete 'megafauna fruit', evolved for dispersal by extinct elephant-like creatures called gomphotheres. Not surprisingly, with no living animals to disperse their seeds, these trees are now themselves becoming endangered. In Madagascar many plants grow thin zig-zag branches to protect themselves from leaf-munching elephant birds, another giant flightless bird that became a casualty of *Homo sapiens* – and that laid eggs so large it is thought to have inspired the legend of the roc in *Sinbad the Sailor*. Modern-day Siberia's wet peaty tundra may stem from the loss of the mammoths, whose earlier grazing nourished a much more productive dry steppe-type biome before their extinction at human hands a mere 2,000 years ago.[18] In Africa elephants play a key role in opening up forests by pushing over trees – a function their relatives in the Americas would also have served before being wiped out by man. In all cases, the vanished megafauna maintained a more diverse ecosystem than the simplified one that replaced them after their sudden demise.

All told, the Quaternary Megafaunal Extinction between 50,000 and 3,000 years ago carried off about a half of the world's large animals (including 178 species of large mammals). This was an extinction wave that bears comparison with the largest in the geological record – but it is still only a prelude to what was to come. The wipe-out that accompanied human migration across the continents was restricted only to the most large-bodied and easily targeted species. In comparison, today not only are the largest animals still at risk, but also small amphibians, songbirds, flowering plants, insects and much else besides. The Sixth Mass Extinction, or the Anthropocene Mass Extinction, is already well advanced – and the death toll will soon rival that at the end of the Cretaceous, when the dinosaurs (and half of the rest of life on Earth) disappeared. Today the small as well as the large wait in line for the cull.

THE SAD STORY OF THE SEA

Perhaps the ecosystem that has been most depleted of its animals in the modern era is the least visible one: the sea. Whilst disappearances on land are comparatively easily studied and recorded, what goes on beneath the waves is an enduring mystery, and humans have traditionally – and tragically – viewed the sea's bounty as limitless. History once again provides a cautionary tale: the whaling industry, for example, managed to reduce cetacean populations once in the hundreds of millions to near-extinction in just a couple of centuries. The sheer scale of the effort was enormous: in the mid-nineteenth century, when many Atlantic whale species had already been exterminated, some 650 whaling ships operated in the Pacific, employing 13,500 seamen.[19] Southern right whales saw their population reduced to as few as 25 breeding females by 1925,[20] after nearly two centuries of devastating slaughter: a low-end estimate is that 150,000 were killed between 1770 and 1900.

Today the eastern North Atlantic right whales are marked as 'critically endangered, possibly extinct' on the IUCN Red List, whilst in the western Atlantic a population of about 300 individuals qualifies merely for 'endangered' status.[21] Several are still killed each year by collisions with ships and through entanglement in fishing nets. As each species was destroyed in turn in its primary areas, the industry moved further afield, killing whales from Antarctica to the Galapagos Islands. Calving grounds were often targeted: congregating mothers could be killed while at their most vulnerable and calves captured too or left to starve. Each population was exploited to near-extinction. Most whales are slow-breeding, and with reproduction rates of 1–3 per cent per year the economically rational whaler would gain more benefit from driving the species to extinction and investing the profits elsewhere (to accumulate interest at perhaps 5 per cent a year) than leaving any alive in the sea.[22] Such is the remorseless logic governing the unregulated capitalist exploitation of nature.

As technology improved, so the slaughter worsened. Steam ships could pursue and kill the fastest species, whilst factory ships could process carcasses at sea without having to call at a port. One after the other, blue, sei, fin, humpback, sperm and minke whales were wiped out over most of the ocean. New whaling grounds would be exhausted at most after a decade, sometimes from one year to the next. All told, the twentieth century saw the slaughter of about 3 million whales, leaving only between 10,000 and 25,000 blue whales in the whole world. The killing goes on still, thanks to the 'scientific whaling' loophole (more like a chasm) in the current International Whaling Commission (IWC) system. Norway, Iceland and Japan continue to kill whales today using the fig-leaf of scientific research, and these countries and their allies have recently tried to overturn the whaling moratorium altogether at the IWC. Whilst it is plausible that stocks of smaller whales like minkes can support a sustainable annual catch, there is a stronger case for leaving the whales alone altogether until their numbers – and the marine ecosystem generally – can properly recover.

Although no whale species were driven to outright extinction, some marine animals have been extinguished completely. The Steller's sea cow, a gentle and intensely social Pacific species, was wiped out for its meat and blubber in the mid-eighteenth century. The great auk – a flightless penguin-like seabird that once lived in huge numbers around the North Atlantic – was also exterminated in a determined campaign of slaughter. Once clubbed to death, the bodies would be plunged into boiling water, their feathers torn out (for stuffing pillows and mattresses, as well as adorning hats), whilst the carcass would be boiled for its oil (used for lighting lamps) and the remainder used to fuel the fires that powered the whole ghastly enterprise.[23] Ship crews would move onto remote islands with the sole purpose of killing as many birds as possible during the summer months. Even on the brink of extinction, the hunting continued: the last breeding pair of great auks were beaten to death in Iceland on 3 June 1844, and their single remaining egg was broken.[24]

Early seafarers were not exactly sentimental about the creatures they encountered. William Dampier, writing about the fur seals he saw on Juan Fernandez island in 1709, marvelled at their beauty, agility and grace, 'how they lie at the top of the water playing and sunning themselves' as he put it. But like everyone, Dampier soon got down to business. 'A blow on the nose soon kills them,' he added helpfully. 'Large ships might here load themselves with seal-skins and Trane-oyl [oil]; for they are extraordinary fat.'[25] And large ships did just that, reducing the island's enormous colonies of seals down to an eventual grand total of just two hundred individuals. One American naval captain related in 1891 how the shooting of fur seal females at sea left their offspring on the shore to starve: 'Thousands of dead and dying pups were scattered over the rookeries, while the shorelines were lined with emaciated, hungry little fellows, with their eyes turned towards the sea uttering plaintive cries for their mothers, which were destined never to return.'[26]

Species after species was relentlessly pursued. Walruses were boiled down for their oil. Giant tortoises were seized in raids on the Galapagos Islands and kept alive by being turned on their backs in ships' holds for months at a time before being eaten for their meat. In 'one of the great wildlife exterminations of colonial times', as marine historian Callum Roberts puts it, an original population of 50–100 million hawksbill turtles in the Caribbean was reduced to just a few thousand (it is still critically endangered worldwide).[27] Sea otters, which once swam in their millions in Pacific coastal waters from Mexico to the Arctic, were reduced to fewer than two thousand by 1911. As industrialisation proceeded, the depletion of whole areas could speed up: when seal colonies were first discovered in the remote South Shetland islands in 1820, a quarter of a million were killed and the population brought to near-extinction within just three years.[28]

All this is in the past, of course. But its impacts are still very much with us, and in many different ways the global slaughter continues. There are no large wild animals left on our planet in anything like the abundance they once enjoyed. Those few hunted species that remain are still under intense pressure; it is as if humanity has learned nothing from past exterminations. Today the extinction of the bluefin tuna is an imminent threat: quotas set at the time of writing by the sadly misnamed International Commission for the Conservation of Atlantic Tunas are high enough to permit fishing boats to catch every single adult bluefin during next year's season.[29] The fish don't have much of a sporting chance: illegal spotter planes guide industrial fleets to wherever the last few thousand individuals can be found.[30] Nor have the economics changed much since the days of whaling: the trading conglomerate Mitsubishi was recently accused of stockpiling frozen bluefin in expectation of a post-extinction price bonanza.[31] With individual fish worth up to $100,000 on the Tokyo sushi market, the tragedy of the commons plays out anew every time the tuna fleets set sail.

The destruction of fish habitat is also routinely ignored in the interests of short-term profit. The North Sea off England's east coast, for example, was not always the murky and uninviting body of water it is today: once its waters were kept clean and sediment-free by rich oyster beds on the sea floor – but these have been ploughed up by trawlers and the sea bottom reduced to a muddy, turgid wasteland. The pressure is unrelenting: intensively fished areas can be hit tens of times in a single year. Deep cold-water corals thousands of years old, supporting flourishing colonies of other marine life, can be reduced to rubble by a single pass of a trawler. Photographs of trawled coral colonies show piles of stony wreckage like the ruins of a pillaged city.

Oceanic island birds are some of the most threatened species anywhere because they are particularly vulnerable to predation by introduced alien invaders. Half of Hawaii's 140 native bird species are now extinct, thanks to the devastation wrought by introduced rats, pigs and cats. On Australia's Christmas Island, the Pipistrelle bat population (I realise bats are mammals, but the point is the same) has plummeted by 90 per cent in the last decade (down to a mere 250 mature individuals), due largely to predation by invasive species like wolf snakes, rats and feral cats.

Consequently, one of the quickest wins for biodiversity conservation is the elimination of alien species from islands. In the biodiversity 'hotspot' of the Galapagos Islands, 140,000 marauding goats have been removed, whilst in the islands off western Mexico – well-known for their unique species and thriving seabird colonies – cats, rats, goats, pigs, donkeys and rabbits have all been removed to protect endemic animals and plants from destruction. The cost has been tiny, compared with the benefits achieved: just $20,000 per colony for 200 seabird colonies protected, and $50,000 per species for 88 endemic species that are found nowhere else on Earth.[32] That any species anywhere else might be lost for the want of such paltry sums would be a terrible indictment of our current lack of concern for the myriad of plants and animals that share this planet with us.

BIODIVERSITY AND THE EARTH SYSTEM

Of course, we may fret about biodiversity loss, but life in general is incredibly resilient. Living species have colonised every nook and cranny of the planetary system. Spiders, anchored by tiny threads, whizz across the stratosphere carried by hundred-mile-an-hour jet-stream blasts. Thermophilic bacteria cluster hungrily around deep-sea volcanic fissures where temperatures soar well past boiling point. Oil-well samples show flourishing microbial life 2 kilometres or more below our feet.[33]

Extraordinary diversity is everywhere: a single 30 g sample of soil from a Norwegian forest has been estimated to contain 20,000 different species of bacteria.[34] We are ourselves walking ecosystems: tiny mites crawl around in our eyelashes, whilst billions of bacteria populate our guts. Higher forms of life may be fewer in number, but are far more varied in form. All told, there are estimated to be 11 million species in the world – with countless more waiting to be discovered. Scientists working on a 2009 update for a global biodiversity report first issued in 2006 had to add 48 new reptiles, 200 new fish and 1,184 flowering plants, all identified for the first time in the intervening three years.[35] Recently ecologists working in the crater of a single extinct Papua New Guinean volcano found 16 new frogs, three new fish, a giant bat and giant rat; luckily a BBC camera crew was on hand to record each unique moment of discovery.[36]

But who cares anyway? Here's Marcel Berlins, columnist on the *Guardian*: 'I passionately believe in saving the whale, the tiger, the orang-utan, the sea turtle and many other specifically identified species. What I do not accept is the general principle that all species alive today should carry on existing forever. We have become so attuned to treating every diminution of animals, insects, birds or fish with concern that we have forgotten to explain why we think it so terrible.' Warming to his argument, Berlins concludes: 'How many mammal species can you think of? Can the remainder be that important? Can their loss matter that much, to you or to the world? Of course we must fight hard to retain as many species as we can; but it isn't a tragedy if we lose quite a few along the way.'

Berlins's common-sense argument is a reasonable one, and its answer not as obvious as one might expect. After all, the biosphere has lost woolly mammoths, Tasmanian tigers and countless other charismatic species already, and yet the world goes on turning. Environments we previously assumed were pristine, like the Amazonian rainforest or the Siberian tundra, now turn out to be more of a product of human engineering than we once thought – and their vanished megafauna have left little identifiable trace, and certainly not one that affects our current lives from day to day. Indeed, most people are unaware that the Quaternary Megafaunal Extinction even happened, and view the disappearance of the mammoth as an interesting but still unsolved mystery, if they think about it at all. Does it really matter if the thinning-out process accelerates a little more?

There are some good utilitarian arguments to show why destroying biodiversity is not a good idea. The biologist E. O. Wilson tells a story of how a small tree in a remote swamp forest in Borneo yielded an effective drug against HIV – except that when collectors returned to the same spot a second time they found the tree had been cut down, and no more could be found.[37] (Happily for AIDS sufferers, a few remaining specimens were eventually located in the Singapore Botanic Garden.) Who knows which tangled Amazonian vine might one day deliver a cure for cancer? But this is only part of the story, for it is ecosystems in their entirety that are valuable and irreplaceable as much as the individual species they contain. Biodiversity loss is a planetary boundary of the utmost importance not because killing off species is morally wrong, but because a healthy diversity of living organisms is essential for ecosystems to function properly.

Living systems keep the air breathable and water drinkable for themselves and us, but to continue to perform these vital services they need to retain their complexity, diversity and resilience. Once humans start to pick off component parts, an ecosystem may appear to function as normal for a while – until some unpredictable tipping point is reached, and collapse occurs. Conceptually this is a bit like the game of Jenga, where wooden blocks are built together in a tower and pieces removed from underneath one by one by each player. Needless to say, whoever removes the crucial 'keystone' piece that topples the tower loses. The lesson of Jenga is an important one, because it shows that there is no single keystone: each removed block makes the tower less and less stable, but no one knows in advance which piece will lead the tower to collapse.

Keystone predators are particularly important to ecosystems. In the marine realm, great sharks – like tiger, hammerhead, bull and thresher sharks – have in recent years been mercilessly targeted worldwide: their numbers have plunged by up to 99.99 per cent in some seas.[38] On the eastern North American coast, rays are no longer being eaten by the vanished sharks, and have increased their numbers as a result. They in turn eat scallops and oysters, destroying the formerly productive scallop fishery.[39] The process is known as a 'trophic cascade' and is now understood to be a fundamental part of ecological dynamics. An ecosystem shift can be irreversible: the Newfoundland cod, whose numbers collapsed because of overfishing in 1992, are unlikely ever to return in substancial numbers. Cod larvae are eaten by smaller fish and crustaceans like lobsters (once kept in check by more numerous adult cod), which dominate the ecosystem instead.[40]

For land-based ecosystems apex predators are just as important. In Yellowstone, the reintroduction of wolves in 1995 has allowed the regrowth of native aspen trees for the first time in half a century. This is because elk populations are now being controlled by wolf predation, preventing overgrazing and allowing trees to recover.[41] In nearby Grand Teton National Park in Wyoming small birds like the gray catbird and MacGillivray's warblers may depend for their survival on wolves, recently reintroduced to the area after an absence of 75 years. Both birds flourish in riverside willows: but the willows, like Yellowstone's aspens, were being overgrazed by hungry moose. In places where predators are still absent, expensive management schemes have to artificially keep down the populations of deer and other grazing herbivores – a service that wolves perform for free.

However, it is not only predators that count. Bottom-up interference can also dramatically destabilise an ecosystem. In the early 1980s a new pathogen appeared in the Caribbean near the mouth of the Panama Canal, wiping out sea urchin populations with extraordinary virulence: within a year 98 per cent of the urchin population was gone, in what is still the worst recorded die-off of any marine animal in history. Because urchins are herbivorous grazers they perform an important function on reefs, keeping the corals clear of algae and seaweed that would otherwise choke the reef systems. Without them, the corals lacked protection, and within a year reefs from Jamaica to the coast of Venezuela disappeared under a thick layer of green slime.[42] After a decade, just 5–10 per cent of the original coral cover was left,[43] and little more remains to this day.[44] A whole marine ecosystem had irreversibly collapsed because of the removal of one of its key components.

Functioning ecosystems need not just a varied number of species, but also – just as crucially – habitat. Humans have disturbed, fragmented or ploughed up huge areas of the planet's terrestrial surface.

But there is a direct correlation between biodiversity and land area: the smaller the remaining fragment, the fewer species it can support. This so-called 'species–area relationship' was illustrated by a massive – though unintentional – field experiment beginning in 1986, when a gigantic hydroelectric dam was built in the jungles of Venezuela. When the lake behind the dam began to fill, the rising tide turned a hilly area of four thousand square kilometres into isolated islands, each with its tropical forest plant and animal species cut off by the surrounding waters. Some of the new islands were very small, just an acre or two in size, whilst others were relatively large, with areas of 150 hectares or more. As you might expect, the smallest islands lost the most biodiversity – three quarters of their original complement – due to their small areas. All islands, large and small, lost their top predators: the jaguar, puma and harpy eagle. But the species that did survive quickly became more abundant as both competition for food and predation ceased abruptly. Some islands were overrun by leaf-cutting ants. One, having housed a large herd of capybaras as the waters rose, ended up as little more than bare ground covered by capybara dung. On some islands, monkeys decimated bird populations, whilst on others rodent populations increased 35-fold.[45] In all cases, complex and formerly diverse ecosystems were torn apart and thrown into chaos.

From these and many other examples, ecologists now understand a fundamental principle of biodiversity: that the greater the diversity of species, the more resilient and stable an ecosystem can be. The same, of course, applies to the biosphere as a whole. We are only just beginning to realise all the myriad ways that different species act unconsciously together to keep this planet habitable and its climate tolerable. Might there be some kind of global 'tipping point' – like the ones that were passed in the Newfoundland cod fishery and the Caribbean coral reefs – where some kind of irreversible global ecosystem shift takes place? This is the possibility that the planetary boundary on biodiversity is intended to prevent: it is now absolutely clear that the Earth's living biosphere depends fundamentally on the maintenance of a broad level of species diversity. If the Sixth Mass Extinction is allowed to continue – or still worse, accelerate further

– then the chance of a global-scale ecosystem collapse can only continue to grow.

THE PRICE OF PANDAS

The current crisis in biodiversity tells us loud and clear that conventional approaches to conservation have failed. 'Paper parks' – named but barely protected – in developing countries are routinely violated by poachers and loggers. What areas are set aside for nature reserves are too small and too fragmented. At sea fishermen compete with each other in a global race to the bottom, knowing that if they do not catch the last bluefin tuna, someone else will. No wonder the 2010 Global Biodiversity Outlook report is full of ominous words and phrases like 'serious declines', 'extensive fragmentation and degradation', 'overexploitation' and 'dangerous impacts'. To meet the planetary boundary, we need to make urgent changes in policy.

Biodiversity loss is fundamentally an enormous market failure, because the people that profit from destroying biodiversity are not generally the same people who lose out when the rainforests, mangroves and coral reefs are finally gone. When palm-oil companies move into the last remnants of rainforest in Borneo, the biofuels they sell deliver benefits to shareholders and foreign consumers, but local people are the losers, as are all the rest of us because of the destructive impact on the world's climate and ecosystems. Our chief task today is to design systems that value nature in a direct and marketable sense and deliver hard cash to those who are in a position to protect ecosystems in a reasonably intact state. What is needed is not more moralising, but more money.

This kind of talk makes many environmentalists queasy. Greens generally view biodiversity conservation as a moral cause, and any discussion of financial mechanisms and marketing schemes arouses strong and principled opposition. Why should any other species, each with just as much right to occupy this living Earth as us, be forced to 'pay its way'? This objection is understandable but wrong-headed: what I am proposing is not a liquidation of nature to make money, but using money simply as a convenient means to safeguard its protection. Money is a measure of value: put a price on wild animals and plants and we will put a value on them too. This is a pragmatic strategy, only to be used in desperation because the others have failed.

But how can the value of natural systems be quantified, let alone brought into the market? A possible approach is to try to assign an imputed shadow price to the ecosystem services – fresh water, clean air, recreational benefits and so on – that different habitats deliver. One study suggests a value of $200,700 per square kilometre for 'high-biodiversity wilderness areas', whilst another finds that 'endemic bird areas' might be worth $88,710 per square kilometre.[46] The imputed value of coral reefs – as destinations for tourism, nurseries for commercially valuable fish and shoreline protectors against storms, for example – has ranged from $100,000 to $600,000 per square kilometre.[47] The values of individual species have also been quantified, based on estimates from public surveys of 'willingness to pay' to prevent their elimination. Using this methodology (and in 2005 US dollars) the Eurasian red squirrel is worth $2.87; the California sea otter $36.76; the giant panda $13.81; the Mediterranean monk seal (almost extinct): $17.54; the blue whale: $44.57; the brown hare: $0.00; the Asian elephant: $1.94; the Northern spotted owl: $59.43; and the loggerhead sea turtle: $16.98.[48]

One team of scientists, led by Robert Costanza – a member of the planetary boundaries expert group – even went so far as to publish an aggregate monetary value of the whole biosphere. There is a conceptual flaw in this, as many have pointed out, because the human economy is a subset of the natural biosphere and could not in any conceivable way replace it. As one environmental scientist sniffed: when it comes to pricing the biosphere as a whole, 'there is little that can usefully be done with a serious underestimate of infinity.'[49] Even so, Costanza and colleagues came up with a precise figure for 'the total economic value of the planet' of $33 trillion per year (as compared with a total global GNP of, when the paper was written in 1997, $18 trillion).[50]

The problem with these figures however is not that they are too precise but that they are not real. No one pays anyone else $33 trillion a year to protect the planet from destruction, nor are any of us actually forking out $17.54 to keep Mediterranean monk seals from going extinct. Yet in a globalised capitalist economy actual, real-world revenue flows are essential if they are to compete with the commercial drive that is destroying and displacing the remaining bits of natural ecosystem worldwide. Mangroves may be valuable as protection against storms and shelter for fish, but someone needs to be paid to look after them if they are not to be chopped down to make way for lucrative shrimp farms. In other words, a financial constituency needs to be created that has a vested interest in protecting its assets – assets that are, in this case, natural rather than commercial capital.

The starting point for this process has to be valuing natural capital. As Pavan Sukhdev, lead author of the 2010 *The Economics of Ecosystems & Biodiversity* (*TEEB*) report, is fond of saying: 'You cannot manage what you do not measure.' One of the report's key recommendations is that the present system of national accounts should be 'rapidly upgraded to include the value of changes in natural capital stocks and ecosystem service flows'. The *TEEB* report consciously encourages the use of banking and accounting terminology with regard to biodiversity: its authors have launched a 'Bank of Natural Capital' website to encourage wider awareness of the ideas it raises. This even extends to proposing an 'internal rate of return' for ecosystems, which varies from 40 per cent for woodlands to 50 per cent for tropical forests to 79 per cent for better-managed grasslands.[51] 'The flows of ecosystem services can be seen as the "dividend" that society receives from natural capital,' the *TEEB* Synthesis Report suggests.[52]

If this all sounds rather capitalistic, it is worth noting that the biggest losers from the current largely unregulated and unquantified degradation of natural capital are the world's poor. The *TEEB* report stresses that forests and other natural ecosystems make an enormous contribution to the so-called 'GDP of the poor' (up to 90 per cent) and that conservation efforts can therefore directly contribute to poverty reduction. In contrast, one estimate of the 'environmental externalities' (the off-balance sheet costs offloaded onto the environment) of the world's top 3,000 listed companies totals around $2.2 trillion annually.[53] All of this value is going into the pockets of corporate shareholders, where it is unlikely to benefit the poor. Moreover, insisting that natural systems are priceless, as many campaigners do, is in practice akin to setting their effective price at zero. The language and practices of economics may offer the strongest tools today for use in nature conservation.

But these imputed values need to be translated into real monetary worth if the natural assets that generate them are to be properly protected. One of the most promising ways of doing this is known as 'payments for ecosystem services' – designing revenue streams that go to communities and landowners who need to be persuaded to keep wetlands and forests intact. In Mexico the annual rate of deforestation has been halved since a 2003 law allowed a portion of water charges to be paid out to landowners willing to preserve forest lands and reduce agricultural clearances. So far 1,800 square kilometres of forest have been protected at a cost of $300 million, both safeguarding biodiversity and reducing greenhouse gas emissions to the tune of 3.2 million tonnes.[54] In the Maldives, whose government I work for as an environmental adviser, one of the schemes under consideration is a levy on diving trips to fund the creation and policing of marine parks. Thus those who benefit from biodiversity – the foreign tourists who marvel at the reef sharks, manta rays and myriad of brightly coloured reef fish that swim around Maldivian coral atolls – can be asked to pay to conserve it.

In other countries, 'biodiversity credits' are being designed that might offer a revenue stream rewarding those who protect and manage biodiverse habitats. In New South Wales, the state government's environment department has set up a 'BioBanking' scheme where developers and landowners can trade biodiversity offsets. Some private companies have been making similar pioneering moves: in Borneo the local government has partnered with the Australian company New Forests to provide an income for the protection of its 34,000-hectare Malua Forest Reserve. Both individuals and businesses can purchase 'Biodiversity Conservation Certificates' that represent the 'biodiversity benefits of 100 square metres of protection and restoration of the Malua Forest Reserve' – habitat for 'endangered wild orangutans as well as gibbons, clouded leopards, pygmy elephants, and over 300 species of birds', according to the Malua BioBank website.[55]

As with carbon offsets, aimed at mopping up an equivalent amount of greenhouse gases to those unavoidably released elsewhere, a partnership between businesses, governments and conservationist groups is currently developing the concept of biodiversity offsets. Their goal is to design offsets that compensate for biodiversity impacts arising from business activities like mining and dam-building, potentially raising considerable sums to protect and enhance ecosystems elsewhere. To count as offsets, schemes must be additional to what would otherwise have happened, provide benefits that last as long as the damage they are intended to address, and deliver equitable outcomes that bring benefits to local people and communities. In addition, offsets are recognised as only being appropriate as a last resort: the so-called 'mitigation hierarchy', in order of importance, is avoid, minimise, restore, and only then offset.[56] Like achieving carbon neutrality, the principle of 'no net loss' of biodiversity – or even better, 'net positive impact' – should and hopefully soon will become part of mainstream business practice.

Protecting natural systems can provide value for money even in the most direct sense. Creating marine protected areas enhances fish stocks, providing benefits both to biodiversity and fishermen in neighbouring areas. The World Bank and UN Food and Agriculture Organisation have estimated that $50 billion is lost each year in terms of economic benefits that could be realised if the world's fisheries were managed sustainably.[57] It may seem counter-intuitive, but a reduction of fishing effort could lead to an increase in overall fish catch. This is a matter of life and death for the over 1 billion mainly poor people who are dependent on fish for their primary source of protein, and whose coastal fisheries have often been scoured out by foreign trawlers from rich nations whose own seas are exhausted.

But voluntary measures will only achieve so much. For biodiversity protection to really work, and for the funds to flow, it needs to be given the force of law. Here too recent progress gives cause for some qualified optimism. The Convention on Biological Diversity, long the poor relation of the Convention on Climate Change, enjoyed a boost in October 2010 with the agreement by world governments of a 'Strategic Plan' for the decade to 2020, intriguingly subtitled 'Living in harmony with nature'. The Plan directs governments to mainstream biodiversity concerns 'throughout government and society', and to take 'direct action … to restore biodiversity and ecosystem services' by 'means of protected areas, habitat restoration, species recovery programmes and other targeted conservation interventions'.[58] These requests are still voluntary at the international level, but national governments are encouraged to turn them into law to ensure that companies, individuals and institutions take biodiversity seriously.

Perhaps just as importantly, a new scientific body is being established, aiming to provide the same expert advice on biodiversity as the IPCC does on climate change. The Intergovernmental Science-Policy Platform on Biodiversity and Ecosystem Services (IPBES) could help finally put this issue at the top of the international scientific and policy agenda, compiling data and producing landmark reports that can inform the efforts of governments and other policymakers.

Biodiversity is an issue whose time has come. All we need to do now is figure out how to pay for it. Remember, all it will cost to save the tiger from extinction is a mere $82 million a year. Rather than passively lamenting its demise, we need to roll up our sleeves and start raising funds. If you do only one thing after reading this chapter, join this effort today.

CHAPTER THREE

The Climate Change Boundary

That climate change is a planetary boundary will come as a surprise to no one. What may come as a surprise however is that the target that has been advocated by not just governments, but environmentalists too, has for years been much too weak. More recently that has begun to change: now an extraordinary coalition of more than a hundred governments and dozens of campaigning groups is lining up squarely behind a safe target for carbon dioxide in the atmosphere, as proposed by the planetary boundaries expert group. Although powerful countries like the US and China are a long way from endorsing this target – and the world economy is even further away from meeting it – the fact that such a crucial planetary boundary has attracted such a strong level of support is a serious piece of good news and one that deserves celebration.

Previous chapters explained how humanity has risen to global prominence through a massive exploitation of fossil energy resources. Human civilisation remains over 80 per cent dependent on fossil fuels worldwide, and as the economy grows so does the rate at which the carbon dioxide resulting from the burning of coal, oil and gas accumulates in the air. On average the carbon dioxide concentration of the atmosphere rises by about 2 parts per million (ppm) every year, from a pre-industrial level of 278 ppm to about 390 ppm today. Whilst the precise level of temperature rise implied by higher CO_2 is always going to be uncertain, it is indisputable that – all other things being equal – global warming will result from the human emission of billions of tonnes of greenhouse gases, sustained over more than a century.

Arguments over what would be a 'safe' level of atmospheric CO_2 have raged for decades. Back in 1992 the UN Framework Convention on Climate Change required in its much-cited Article 2 that the objective of international policy should be to avoid 'dangerous anthropogenic interference' in the climate system – but without defining what 'dangerous' actually meant. The British government's *Stern Review on the Economics of Climate Change* of 2006 suggested a stabilisation target of 550 ppm CO_2e (carbon dioxide-equivalent, implying a bundling together of all climate-changing gases rather than only CO_2). Two years earlier, the European Union had endorsed a target of limiting temperature rises to 2 degrees Celsius, implying – it was stated – a CO_2 target of 450 ppm. This latter objective was endorsed in my 2007 book about climate-change impacts, *Six Degrees*, where I suggested that 2 degrees and 450 ppm were necessary to steer away from large-scale dangerous tipping points in the climate system. Major environmental groups also lined up behind similar targets, and pushed them hard at international meetings.

It turns out we were all wrong. A fair reading of the science today, as this chapter will show, points strongly towards a climate change planetary boundary of not 450 ppm but 350 ppm for carbon dioxide concentrations – a level that was passed back in 1988, the year that NASA climate scientist and planetary boundaries expert group member James Hansen first testified to the US Congress that global warming was both real and already under way. Hansen has done more than any other scientist to put the 350 number on the map. He was one of the first to realise its importance, and has become a tireless advocate of the actions that are necessary to meet it. It was Hansen's discussions with the American author and activist Bill McKibben, indeed, that led to the creation of the worldwide movement 350.org. McKibben calls 350 'the most important number in the world', and he is right.

Never mind the enduring global-warming controversies in the media; these are a distraction. The climate change planetary boundary is the one that is best understood, and that we know most about how to achieve. Moreover, meeting the boundary is a basic requirement for any level of sustainable planetary management: if CO_2

continues to rise, and temperatures begin to race out of control, then the biodiversity boundary, the ozone boundary, the freshwater boundary, the land use boundary and ocean acidification boundaries cannot be met either, and the remaining planetary boundaries are also called into question.

The climate boundary is humanity's first and biggest test that will reveal early on whether we are truly capable of managing our environmental impacts in a way that protects the capacity of the biosphere to continue to operate as a self-regulating system. It is a testament to our intelligence that we have developed our scientific understanding so far that we now know a great deal about how the climate system works, and can define with some confidence where the planetary boundary should lie. It is perhaps testament to our stupidity, however, that despite decades of research and advocacy on climate, all pointing at the need to control greenhouse gas production, human emissions today continue inexorably to rise.

Thankfully the technologies and strategies that humanity needs to achieve the climate boundary are today no mystery. We have all the tools necessary to begin a wide-scale decarbonisation of the global economy, and to achieve this at the same time as both living standards and population numbers are rising rapidly in the developing world. But environmentalism will need to change at the same time. Much of what environmentalists are calling for will either not help much or is actually thwarting progress towards solving climate change. It is time for a new – and far more pragmatic – approach, that does not hold climate change hostage to a rigid ideology.

350: CURRENT EVIDENCE

First we need to establish whether 350 is actually the right number, and one that is supported by science. There are three broad lines of evidence that support the conclusion that atmospheric CO_2 concentrations need to be limited to 350 ppm. The first is the sheer rapidity of changes already under way in the Earth system, changes I never dreamt I would see so quickly when I started working on this subject more than ten years ago. These warn of looming danger. The second is modelling work suggesting that positive feedbacks – or thresholds, or tipping points, call them what you like – are getting perilously close. The third, and perhaps most conclusive, is evidence from the distant past linking temperatures with carbon dioxide concentrations in earlier geological epochs.

The best place to look for confirmation that our planet is gaining heat is not the air temperature at the ground, but the energy imbalance – the difference between incoming and outgoing radiation – at the very top of the atmosphere. There our sentinel machines, the satellites silently orbiting the planet twenty-four hours a day, show clearly that outgoing longwave heat radiation is increasingly being trapped at exactly those parts of the spectrum that correspond with the different greenhouse gases building up in the atmosphere below.[1] Natural variability is important in determining the average temperature each year, but recent records are revealing: the hottest year on record, according to NASA, is now tied between 2010 and 2005, with 2007 and 2009 statistically tied for second- and third-hottest.[2] Whatever the individual temperature records, the climatic baseline is visibly shifting: every year in the 1990s was warmer than the average of the 1980s, every year of the 2000s warmer than the 1990s average.[3]

There are now multiple lines of evidence pointing to ongoing global warming, some of which show that we are altering the characteristics of the atmosphere in unanticipated ways. Air-pressure distribution is changing around the world, with rises in the subtropics and falls over the poles.[4] The stratosphere has cooled as more heat is trapped by the troposphere underneath,[5] whilst the boundary between these two higher and lower atmospheric layers has itself increased in height.[6] Even the position of the tropical zones has begun to shift as the atmosphere circulates differently in response to rising heat.[7]

A more energetic atmosphere also means more extreme rainfall events as the levels of water vapour in a warmer atmosphere increase: this too has been observed.[8] The catastrophic flooding events that hit Pakistan in August 2010 and Australia in January 2011 are exactly the kind of hydrological disasters that will be striking with deadly effect more often in a warmer world. Whilst people in poorer countries are most vulnerable to the effects of floods, any country can be hit at any time: in the English Lake District the heavy rainfall event of 18–20 November 2009 had no precedent: rainfall totals outstripped previous all-time records in over 150 years of measurements.[9]

Perhaps the clearest indicator of current danger – Ground Zero for global warming – is the rapid thaw of the Arctic. Few experts argue any more about whether the sea ice sheet covering the North Pole will melt completely; merely when. In recent years the Arctic ice cap has entered what Mark Serreze, a climatologist at the National Snow and Ice Data Center (NSIDC) in Boulder, Colorado, calls a 'death spiral'.[10] The extent of Arctic ice is plummeting, and what remains is thinner and more vulnerable to melt than before. In terms of volume, less than half the ice cap of the pre-1980 era remains; more than 40 per cent of the volume of multi-year ice (the thicker stuff that lasts through the summer) has disappeared since only 2005.[11] Even the wintertime ice coverage is in decline: in January 2011 the NSIDC announced that the sea ice extent for that month was the lowest in the satellite record, with the Labrador Sea and much of western Greenland's coast remaining completely unfrozen.[12] The year of what I call A-Day, the late-summer day at some time in the future when not a fleck of the North Polar floating ice remains, has been suggested by one modelling study as likely to arrive in 2037, but if recent years are anything to go by this could shift closer by as much as a decade.[13]

A-Day will be a momentous date for the Earth, for it will be the first time in at least five thousand years that the Arctic Ocean has been without any summertime sea ice.[14] This will in turn alter the heat balance of the planet and the circulation of the atmosphere: without its shiny cap of frigid ice, the Arctic Ocean can absorb a lot more solar heat in summer and release much more in winter, changing storm tracks and weather patterns. The resulting prognosis is not for straightforward warming everywhere: one model projection by scientists working in Germany, published in November 2010, suggested that disappearing sea ice in the Arctic Ocean north of Scandinavia and Siberia could in fact drive colder winters in Europe. The researchers proposed that warmer unfrozen waters in the north could drive a change in wind patterns that allows cold easterly winds to sweep down into Europe and Russia, and that this may have helped cause the colder winters of 2005–6, 2009–10 and 2010–11 in both Europe and eastern North America, which have seen snowstorms and frosts even as the Arctic basked in unprecedented winter warmth. 'Our results imply that several recent severe winters do not conflict [with] the global warming picture but rather supplement it,' they concluded in the *Journal of Geophysical Research*.[15]

The disappearance of the Arctic ice will eliminate an entire marine ecosystem. Currently algae growing on the underside of floating ice are the base of a unique food chain, feeding zooplankton that in turn support large populations of Arctic cod.[16] Rapidly diminishing ice spells disaster for ice-dependent species like ringed seals, walrus, beluga whales and, of course, polar bears. This may not necessarily mean outright extinction for the latter, but it will lead at best to a substantial reduction in their habitat.[17] In May 2008 the polar bear was listed as 'threatened' under the US Endangered Species Act thanks to climate change.[18]

Given its current rate of precipitous decline, there is little hope that the Arctic ice cap's death spiral can be arrested. But it is theoretically still possible to save or restore the frozen North Pole – by urgently retreating back within the 350 ppm climate boundary, and, as I will set out in a future chapter, by reducing emissions of other warming agents like black carbon. As NASA's James Hansen, a member of the planetary boundaries expert group, writes: 'Stabilisation of Arctic sea ice cover requires, to first approximation, restoration of planetary energy balance.'[19] Reducing carbon dioxide levels to between 325 and 355 ppm would achieve this initial outcome, Hansen suggests – however, a further reduction, with CO_2 down between 300 and 325 ppm, 'may be needed to restore sea ice to its area of 25 years ago'.

Serious climate impacts have of course also been identified outside the polar regions. In a June 2010 piece for *Science* magazine, climate experts Jonathan Overpeck and Bradley Udall – based at the universities of Arizona and Colorado respectively – wrote that 'it has become impossible to overlook the signs of climate change in western North America'. These signs include 'soaring temperatures, declining late-season snowpack, northward-shifted winter storm tracks, increasing precipitation intensity, the worst drought since measurements began, steep declines in Colorado River reservoir storage, widespread vegetation mortality, and sharp increases in the frequency of large wildfires'.[20] As with the melting of the Arctic, Overpeck and Udall reported that the impacts of global warming in western North America 'seem to be occurring faster than projected' in mainstream climate assessments like the IPCC's 2007 report. In the Rockies higher temperatures mean that more winter precipitation is falling now as rain, and what snow does lie is melting earlier and faster. Peak streamflow in the mountains of the American west now occurs up to a month earlier than it did half a century ago.[21]

One of the most worrying climate impacts mentioned by Overpeck and Udall in the western US is the rapid increase in tree death rates: more than a million hectares of piñon pine died recently due to drought and warming, and even desert-adapted species, that should be able to cope with ordinary dry weather, are 'showing signs of widespread drought-induced plant mortality'. This climate-related forest die-off seems to be part of a serious global trend, which has seen widespread tree death observed in places as far apart as Algeria and South Korea, and dramatic reductions of forest cover even in protected areas like national parks.[22] In some cases insect infestations are the immediate cause of the die-offs: in British Columbia beetle outbreaks have killed such extensive areas of boreal forest that experts estimate 270 million tonnes' worth of carbon sink have been eliminated.[23]

All over the world ecosystems face being wiped out as their climatic zones shift rapidly elsewhere – or disappear altogether. Just as polar animals are effectively pushed off the top of the world by the rising heat, so mountain-dwellers are confined to ever-shrinking islands of habitat on the highest peaks. Indeed, what is possibly global warming's first mammal victim – the white lemuroid possum – may already have disappeared from its habitat of just a few isolated mountaintops in tropical Queensland, Australia. 'It was quite depressing going back on the last field trip a couple of weeks ago, going back night after night thinking, "OK, we'll find one tonight,"' biologist Steve Williams told the Australian Broadcasting Corporation. 'But no, we still didn't find any.'[24] In Madagascar, another global biodiversity 'hotspot', mountain-dwelling species are already being displaced uphill, and some species of frog and lizard may now be extinct because of the changing climate.[25]

Thermal stress also affects humans, of course, as increasingly intense and frequent heatwaves scorch our cities. Hundreds died in the August 2010 Moscow heatwave. Tens of thousands (and possibly as many as 70,000 in total[26]) succumbed across continental Europe in the record-breaking summer of 2003. Very hot summers have already become more frequent across the Northern Hemisphere, and the risk of a repeat of the 2003 heat disaster has now doubled, thanks to global warming.[27] According to news reports, 2010 saw Japan endure its hottest-ever summer, whilst all-time heat records were smashed in 17 different countries.[28] Heatwaves have also increased in the Mediterranean region in number, length and intensity, according to the latest studies.[29] This warming and drying trend is repeated across much of the world: in southwestern Australia, for example, rainfall has fallen by a fifth since the 1970s, leading to permanent water shortages in Perth.[30]

All these lines of evidence – of rising temperatures, thawing ice caps, shifting weather patterns and increasingly dangerous impacts – emphasise that limiting CO_2 concentrations at 350 ppm in order to prevent substantial future global warming is the only sensible option. Getting back within this planetary boundary would potentially restore the Arctic to health and prevent the complete thawing of mountain glaciers in the Andes and Himalayas that help sustain freshwater supplies to many millions of people. Limiting the speed and magnitude of the future temperature increase to just a degree and a half this century, the most likely outcome of a 350 ppm pathway, would keep global warming slow enough to allow both natural ecosystems like coral reefs and human societies to adapt to climate change.

350: MODELLING EVIDENCE

Observing the present allows us to extrapolate using educated guesswork towards the future. But perhaps a more scientifically rigorous way to project future climate change is to look at the output of complex computer models that simulate the way the climate operates in incredible detail. Taking months of supercomputer time to crunch all their complex equations, these modelling studies allow scientists to simulate changing conditions on Earth as CO_2 rises, ice melts and temperatures climb inexorably. Although computer models are always going to be an imperfect representation of the real planet we live on, they are the only way to run experiments into the future – other than sitting back and watching what really happens to the Earth, by which time it will be too late to do anything about it.

The point of setting a planetary boundary on climate is to enable humanity to keep on the right side of potential tipping points that could mark dangerous and potentially irreversible shifts in the way the biosphere operates. With that objective in mind, two members of the planetary boundaries expert group, Tim Lenton and Hans Joachim Schellnhuber, were co-authors of a landmark study published in 2008 that tried to identify the different tipping points that might exist in the climate system and get some idea of what level of temperature rise might trigger them.[31] Top of the list was Arctic sea-ice loss. This is because the Arctic melt is self-reinforcing: as ice disappears, its highly reflective surface is replaced by darker sea or land, that absorbs more of the sun's heat, allowing the melt of even more ice. The problem here is that models generally underestimate the observed loss of ice – in other words, what is happening in the real world tends to be worse than anything projected by the models. Given this, the experts concluded, 'a summer ice-loss threshold, if not already passed, may be very close'. Only a 350 ppm target would likely prevent it, corresponding to 0.5 to 2°C future global warming. But even this may not be enough.

Second on the tipping points list came the melting of Greenland's vast ice sheet. Thick enough to raise the global oceans by seven metres if it melted entirely, the stability of Greenland matters hugely to far-away nations like Bangladesh and the Maldives, which face partial or total inundation (in the case of the latter) if it melts because of global warming. So where does the tipping point lie that might doom the Greenland ice cap to eventual destruction? Between just 1 and 2 degrees above today's temperatures, the experts concluded, meaning that a 350 ppm trajectory is once again the least we will need to achieve to protect it. Here too the process could become self-reinforcing. The centre of Greenland is extremely cold because the thickness of the ice sheet means that it extends into high altitude: Greenland's Summit Camp is located 3,200 metres above sea level. But as global warming nibbles away at the edges of this enormous ice body, more of it comes into the lower altitude zone, exposing the ice to higher temperatures and increasing the melt rate. Although eliminating a whole continent's worth of ice will take time, the process could be completed in as little as three centuries, dramatically changing the coastal geography of the planet. Once again, this is a tipping point that humanity would be wise not to trigger.

Greenland is not the only vulnerable polar ice sheet, of course. Third on the list came the West Antarctic Ice Sheet, again of serious concern because – like Greenland – its loss could trigger multi-metre rates of sea-level rise. The West Antarctic also could be subject to a positive feedback process once a serious melt got under way, not just because of the change in altitude but because most of the ice sheet is actually grounded well below today's sea level. As warming waters penetrate underneath the ice mass they could trigger a collapse that would be unstoppable, and would eventually raise global sea levels by another 5 metres. Here we may be on slightly safer ground, as the experts conclude that a global warming of 3–5°C will likely be necessary to lead to complete collapse. So the 350 ppm boundary would appear to be well within the safety margin according to the models.

As with the Arctic sea ice, however, the real world may prove the models of Greenland and the West Antarctic to be overly conservative. The most recent satellite data from the GRACE (Gravity Recovery and Climate Experiment) mission shows a doubling in ice mass lost from both Greenland and Antarctica over the last decade[32] – despite a thickening of Greenland's higher interior where warmer winds have increased snowfall rates. Until recently the massive East Antarctic ice sheet was probably stable, but it too began losing ice in coastal areas after about 2006.[33] In total the Earth's great ice sheets are now shedding a few hundred billion tonnes of ice annually, and sea levels rising by slightly more than 3 mm per year as a result – nearly double the rate for most of the twentieth century.[34] A rise in sea levels by 2100 of somewhere between 60 cm and 1.6 metres is now on the cards,[35] substantially more than was suggested just a few years ago by the IPCC.[36]

A more familiar tipping point was examined next, one that has even been made into a dramatic Hollywood film. In *The Day After Tomorrow*, a sudden ice age is seen flooding and then freezing New York (why is it always New York?) after global warming destabilises the circulation of the Atlantic Ocean. Although the flash-freezing depicted in the movie is thermodynamically impossible, the scenario of a collapsing Atlantic current is not complete science fiction. All the models examined by the expert group led by Tim Lenton showed a tipping point in the North Atlantic where warmer, fresher waters could shut down the circulation pattern that brings comparatively balmy temperatures to the eastern US and high-latitude Western Europe. This shutdown would not trigger a new ice age, but temperatures in these regions could fall for several decades, causing serious impacts on societies and ecosystems alike.[37] Again unlike the Hollywood movie, which showed temperatures dropping in seconds, the full transition towards an Atlantic Ocean circulation shutdown would likely take a century or more. More good news is that avoiding this tipping point is still possible: the scientists conclude from studying their models that a global warming of 3–5°C would be needed to put us in the danger zone, well above the 1.5°C maximum warming implied by our 350 ppm planetary boundary.

Another candidate on the tipping-point list is the Amazonian rainforest. For years now many scientists have warned that global warming could trigger a collapse of the forest if rising temperatures lead to severe drought in western Brazil. This scenario seems even more of a danger given the recent droughts experienced in Amazonia in both

2005 and 2010, where entire river systems in this normally wet forest dried up for hundreds of kilometres. The problem here is that models don't concur: some show a warmer Amazon getting wetter, whilst the most pessimistic forecasts for Amazon die-back are based on the projections of just one model, the HadCM3 model produced by the UK Met Office's Hadley Centre. However, half of the 19 different models examined by a team of scientists led by Oxford University's Yadvinder Mahli in 2009 did show a shift towards more seasonal forest, and a quarter showed that the rainforest could dry out sufficiently to collapse into a savannah-type ecosystem instead.[38] Keeping global temperatures below 3°C – very likely if our 350 ppm planetary boundary is achieved – should be enough to avoid this transition, but just as important will be respecting the other planetary boundaries on land use and biodiversity loss. The Amazon rainforest today is probably more threatened by deforestation and agriculture than it is by rising temperatures.

If the Amazon rainforest did collapse, huge quantities of carbon would be released in the process, giving a further boost to global warming. But the biggest carbon stores of all lie not in the tropics, but in the sub-polar continental regions where frozen permafrost holds enormous carbon stores tens of metres thick in Siberia and other high-latitude land areas. The threat to permafrost stability is possibly global warming's biggest tipping point, because if this frozen carbon store begins to thaw, vast quantities of both carbon dioxide and methane will be released. According to a 2008 study in the journal *BioScience*, the carbon locked up in the Northern permafrost zone totals more than 1.5 trillion tonnes, double the entire carbon content of the atmosphere.[39] Even if only 10 per cent of this permafrost thaws, another 80 ppm of CO_2 will have accumulated in the atmosphere by 2100, raising the planet's temperature by an additional 0.7 degrees[40] – and making the eventual attainment of the 350 ppm climate change boundary much more difficult.

Scientists have already begun watching with some alarm a recent upward trend in atmospheric methane, some of which may be coming from the Arctic.[41] Not all this methane – a greenhouse gas 25 times more potent than CO_2 – is likely to bubble out of swamps on land;

vastly more is contained in subsea sediments in the form of ice-like methane hydrates. If these hydrates melt rapidly as the oceans warm up, then all global warming bets are off – a scenario that has already sparked scary newspaper headlines. So how afraid should we be? Researchers have already reported seeps of methane leaking from the seabed offshore from eastern Siberia and the Norwegian Arctic islands of Svalbard, in both cases possibly in response to warmer ocean waters.[42] But the experts are cautious. 'Methane sells newspapers, but it's not the big story,' writes David Archer on the excellent RealClimate blog.[43] 'CO_2 is plenty to be frightened of, while methane is frosting on the cake.'

Work by Archer and colleagues modelling the Earth's response to climate change suggests that methane hydrate release could add another half-degree or so to the total warming, but only over several thousand years, and only if the released methane is not dissolved or oxidised first in the ocean before it has time to escape into the atmosphere.[44] This is a 'slow tipping point', Archer concludes: it takes a long time for warming to penetrate the oceans, even longer for this to melt and release hydrates, and longer still for this methane to warm the atmosphere and the oceans further in a positive feedback loop. Happily, this is a tipping point we have still not crossed – 'We have not yet activated strong climate feedbacks from permafrost and CH_4 [methane] hydrates,' reported a team of scientists in 2009.[45] In the case of methane hydrates, respecting the climate boundary is not necessarily about protecting ourselves or even our children, but the stability of the Earth system over the very long term – for this tipping point, while slow to activate, would be essentially irreversible once crossed.

350: PAST EVIDENCE

If current observations of accelerating climate change and worries about tipping points in the future make two very good reasons why 350 ppm is the right place for a climate change planetary boundary, even stronger evidence comes from the Earth's more distant climatic past. Climate models projections such as those published by the IPCC tend to project nice smooth – albeit upward-pointing – curves of likely future temperature trends. But a glance back in time, courtesy of ice-core records drilled in Greenland and Antarctica, shows that gentle, slow changes are far from being the norm in the Earth's past. Instead, these records of past climate – which now reach back almost a million years – show climatic swings of extraordinary and terrifying abruptness. One extremely sudden warming took place in Greenland 11,700 years ago; it involved a temperature rise of 10 degrees Celsius within just three years.[46] Rapid shifts are observed elsewhere too: 12,679 years ago, according to sediments recovered from a lake in western Germany, the European climate saw a sudden transition to more stormy conditions between one year and the next.[47] The lesson is clear. Abrupt climate change is not the exception in the past, it is the norm. As the veteran oceanographer Wally Broecker says: 'The climate is an angry beast, and we are poking it with a stick.'

Although current CO_2 levels are higher than they have been for a million years, if we look even further back into the geological past there are episodes when both carbon dioxide and temperatures were far above where they are now. But rather than suggesting we have nothing to worry about, they further strengthen the evidence for counting 350 ppm as the crucial planetary boundary. For example, during the Pliocene epoch, about 3 million years ago, sea levels were 25 metres higher than today because the major ice sheets were much smaller than now due to a warmer climate. The CO_2 concentration then? About 360 ppm – a line we crossed in 1995.[48]

The Earth was completely ice-free – and sea levels 80 metres or more higher – until about 33 million years ago, early in the geological epoch called the Oligocene. After having been at 1000 ppm or higher throughout the Cretaceous, Eocene and Paleocene, this was the moment when CO_2 levels dropped past a crucial threshold allowing continental-scale ice sheets to form on Antarctica for the first time in perhaps a hundred million years.[49] This CO_2 level was 750 ppm, a level expected to be crossed again in about 2075 if carbon emissions continue to rise unabated. For the following 31 million years, only Antarctica held substantial ice sheets – until, late in the Pliocene, the more recent ice-age cycles began. There was another CO_2 threshold at play here, one that allowed Northern Hemisphere ice sheets (such as the current one on Greenland) to form for the first time. That level was 280 ppm, which we crossed right at the start of the Anthropocene at the turn of the nineteenth century. Were Greenland to be ice-free at the moment, in other words, CO_2 levels are already too high for an ice sheet to form. Once again, 350 ppm seems to be the minimum necessary to protect the big polar ice sheets over the longer term.[50,51]

NASA's James Hansen (a member of the planetary boundaries expert group) wrote in the introduction to his landmark 2008 paper 'Target Atmospheric CO_2: Where Should Humanity Aim?' (published with nine co-authors in the open-source journal *Open Atmospheric Science Journal*): 'If humanity wishes to preserve a planet similar to that on which civilization developed and to which life on Earth is adapted, paleoclimate evidence and ongoing climate change suggest that CO_2 will need to be reduced from its current 385ppm to at most 350ppm, but likely less than that.'[52] Hansen and his colleagues reject a target of 450 ppm, for long the objective of both many governments and environmental groups. 'A CO_2 amount of order 450ppm or larger, if long maintained, would push Earth toward the ice-free state,' they maintain. And although the inertia of the climate system and slow response-times of ice sheets would limit the speed of this change, 'such a CO_2 level likely would cause the passing of climate tipping points and initiate dynamic responses that could be out of humanity's control'.

TOWARDS A TECHNOFIX?

Having said all that, solving climate change is actually a lot simpler than most people think. Global warming is not about overconsumption, morality, ideology or capitalism. It is largely the result of human beings generating energy by burning hydrocarbons and coal. It is, in other words, a technical problem, and it is therefore amenable to a largely technical solution, albeit one driven by politics. I often receive emails telling me that fixing the climate will need a worldwide change in values, a programme of mass education to reduce people's desires to consume, a more equitable distribution of global wealth, 'smashing the power' of transnational corporations or even the abolition of capitalism itself. After having struggled with this for over a decade myself, I am now convinced that these viewpoints – which are subscribed to by perhaps a majority of environmentalists – are wrong. Instead, we can completely deal with climate change within the prevailing economic system. In fact, any other approach is likely doomed to failure.

Here are two options that certainly won't work. First, we could try to reduce the global population. Certainly, fewer people by definition means lower emissions. But getting to 350 ppm by reducing the number of human carbon emitters on the planet is impossible as well as undesirable: at a first approximation it would require the number of people in the world to be reduced by four-fifths down to just a billion souls or less. Short of a programme of mass forced sterilisation and/or genocide, there is no way that this could be completed within the few decades necessary. Certainly there are a multitude of reasons why giving people access to family planning is a good idea, but climate-change mitigation is not among them. The best reason for promoting birth control is that people want it, and everyone should be able to choose how many children they have. The future of the planet doesn't come into it.

The second option is to restrain economic growth, as GDP is very closely tied to the consumption of energy and therefore carbon emissions. No one disputes that recessions do tend to reduce emissions: the global financial and economic crisis that began in 2008 led to a fall in CO_2 emissions worldwide by 1.3 per cent within a year.[53] But imagine that the recession had been caused not by solvency problems within financial institutions but by government policies to tackle climate change. Jobless totals would be rising, government cutbacks in welfare services hitting the poor, and a new age of austerity dawning – all because of the tree-huggers. If you thought the debate on climate change was ill-tempered now, imagine that particular future and its implications.

Greens have for years called into question GDP as a measure of true progress, but the reality is that increasing prosperity – measured in material consumption – is non-negotiable both politically and socially, especially in developing countries. This may one day need to change, but that is a different debate, and one that needs to be had for different reasons. As the climate scientist Roger Pielke Jnr writes in his 2010 book *The Climate Fix*, 'if there is an iron law of climate policy, it is that when policies focused on economic growth confront policies focused on emissions reductions, it is economic growth that will win out every time.' Greens may despair, but I think Pielke Jnr is right. The implication, however, is not that we are all doomed, but that any successful policy to decarbonise the global economy 'must be designed such that economic growth and environmental progress go hand in hand'.[54]

In a related sense, although Greens often insist that energy is too cheap, this too is incorrect. Energy is actually too expensive, certainly for the 1.5 billion poor people in the world who lack access to electricity because they do not have the purchasing power to demand it. Well-fed campaigners in rich countries may fantasise romantically about happy peasants living sustainably in self-reliant African villages, but the fact is that people across the developing world are desperate to increase their economic opportunities, security and wealth. They want to have enough to eat, they want to have clean water and they want their young children not to die of easily treatable diseases – and that is just for starters. They want the benefits of being part of the modern world, in other words, which is why so many young people across the developing world are moving to cities in search of a job and a better way of life. And this better way of life is coming, as the soaring rates of economic growth in China, India, Brazil and many other developing countries demonstrate. The fact is that most of the world needs more growth, not less: China has lifted 300 million people out of poverty in the last couple of decades due to its economic miracle. Hundreds of millions more, in Africa now too as well as Asia and Latin America, are determined to follow, as they have every right to.

By mid-century, in other words, we will see a world of many more, much richer people. Most Greens view this prospect with dread, for how can the world possibly reduce carbon emissions under such a scenario? The London-based New Economics Foundation (NEF), for example, writes in a recent report: 'If everyone in the world lived as people do in Europe, we would need three planets to support us.'[55] This is nonsensical, for everyone in the world is going to live like Europeans within this century (and Europeans too will also get richer) whether NEF likes it or not, and we will still only have one planet. NEF's 'Happy Planet Index' was recently topped by Costa Rica (with the Dominican Republic in second place and Jamaica in third), apparently suggesting that the best country in the world to live is one where 10 per cent of the population still survive on just $2 a day.[56] Certainly, the fact that GDP does not necessarily equate to happiness is an important point to make. But it won't cut much ice with the billions of people – a majority of humanity – who are poor, insecure or malnourished in today's world. For them economic growth is not a choice but a necessity.

So reducing human population and economic growth is neither possible nor desirable. Luckily there is a third way: we can reduce the carbon intensity of the economy, so that for each unit of GDP produced, less and less carbon needs to be emitted. This means deploying low-carbon technologies across the board so that the energy that is needed to drive economic activity can be generated without additional greenhouse gases. What we need, in other words, is an economy-wide technofix.

TECHNOLOGIES FOR 350

My own perspective on tackling climate change has shifted since I was appointed adviser to President Nasheed of the Maldives in 2009. The president, whose country is of course early on the list of those liable to be wiped out by rising sea levels, had just announced his ambition for his nation to become the first carbon-neutral country in the world, by 2020. Suddenly, having spent most of my life as a journalist, I was confronted with the challenges of real energy supply in a real developing country. All my Green ideology – of tackling corporate power, reducing consumption, challenging economic growth and so on – was going to be of little help with this intensely practical challenge. To be carbon-neutral the Maldives would have to stop burning diesel in electric generators on every one of its 300 or so inhabited islands, and shift instead to an energy system entirely based on renewables. It would have to do this in a way that would not raise people's energy bills, and would provide opportunities for new business. I found myself in a world where discussions of wind and solar hybrids, battery storage options, biomass and waste-to-energy, and electrical grid load-balancing came to the fore. I began to think less like an ideologue and more like an engineer.

This, on a far grander scale, is the same challenge that confronts the world. To achieve the planetary boundary of 350 ppm, the global economy needs to be carbon-neutral by mid-century and carbon-negative thereafter. Meeting this target means we all – Greens included – need to start thinking like engineers. This is a huge industrial building project, converting the energy basis of civilisation from fossil fuels to a variety of cleaner sources. If we do it right, it will not be a burden or a cost to the world's economy, but a source of enormous potential future growth, innovation and job creation. The sheer amount of economic activity implied by the transition is staggering: to reduce the emissions of the United States by a third, for example, would (using current technologies) involve constructing 145 nuclear plants, 33,000 solar thermal power stations and 130,000 large wind turbines. In Germany, the same ambition of a 30 per cent emissions cut implies 21 nuclear plants, 4,800 solar stations and 20,000 additional windmills.[57]

Different technologies can be substituted according to different circumstances or national preferences, of course. The Austrians, for example, despise nuclear power. (The country spent $1 billion building a nuclear plant, and then had a referendum in 1978 that was won by the anti-nuclear lobby. The plant, called Zwentendorf, was never opened, and coal-burning power stations built instead.) For the Maldives I would not suggest any nuclear power stations, because each island operates as a separate independent energy entity and nuclear plants are simply too big to be appropriate. Moreover, the country is drenched in solar radiation for most of the year – its main constraint, in fact, is the land space needed to capture the sun's energy. But very large, densely populated nations outside the tropics are likely to need substantial nuclear generation. This may be difficult for many

Greens to swallow, but as I will show in future chapters, nuclear power is nothing like the environmental threat it has long been made out to be. Instead, by displacing coal from our energy mix, it can be a net win for the biosphere. China, for instance, has 13 operational nuclear plants and 150 more under construction or on the drawing board.[58] Each 1-gigawatt nuclear plant will displace 6 million tonnes of annual CO_2 emissions, making this one of the best pieces of climate-related news anywhere in the world.[59] That should be the end of the matter so far as environmentalists are concerned: nuclear is Green.

To cut global emissions in half by 2050 (with growing energy consumption in the meantime) would require the construction of 12,000 nuclear power stations – with one plant coming online every single day between now and then (assuming we start in 2015). I mention this only as an illustrative exercise, for no one – not even the nuclear industry – suggests that we try to deal with climate change using nuclear power only. Such a level of new-build sounds impossible, but consider that over the last fifty years humans have constructed two large dams per day, half of those in only one country – China.[60] France, which still enjoys the lowest per capita emissions in the industrialised world thanks to its near-80 per cent use of nuclear power, managed to commission five reactors per year at its peak of new-build in the 1980s.[61] With 59 operating nuclear reactors, France is the world's largest electricity exporter – much of it to its anti-nuclear neighbour, Germany. According to the Royal Academy of Engineering, the cost of nuclear electricity is roughly comparable to gas, less than coal, and much less than wind, giving the lie to the oft-heard objection that nuclear power is too expensive.[62] By way of comparison, 2010 saw 13 new construction projects begin for nuclear power stations (eight of these in China, two in Russia, two in India and one in Brazil), promising that by 2015 one new plant will be opening every month.[63] The nuclear renaissance is happening, but it needs a step-change to be of any substantial help in tackling the climate planetary boundary.

In March 2011 the earthquake and tsunami in north-eastern Japan set off a crisis at the Fukushima nuclear plant which sent seismic shock-waves around the world's energy industry. Thanks to hyperbolic media coverage about a spreading radioactive cloud, people as far away as China and the United States began panic-buying salt and iodine tablets, whilst several embassies nearer by in Tokyo evacuated their staffs. As I discuss later, the actual dangers from increased radiation were minimal: those leaving Tokyo on aircraft will have soaked up more radiation from cosmic rays on their flights than if they had stayed put in the Japanese capital. No matter: the fear of radiation is vastly more potent than the real thing, and the Fukushima disaster led within days to panicky measures against nuclear power in Germany, China and many other countries. Only time – and a little perspective – will tell whether Fukushima was enough to stall the nuclear renaissance, but I doubt it. Having mothballed several of its nuclear plants unnecessarily, the Germans immediately found themselves more dependent on coal – which by every measure is vastly more toxic and dangerous than nuclear. Japan too has little option but to stick with nuclear: it already imports large quantities of fossil fuels, and with a dense population packed into just a few small islands its renewables options are few.

All-told, if the world moves away from nuclear as a result of Fukushima, several billion more tonnes of carbon will be being emitted annually by 2030 due to the implied resurgence of coal and gas for electricity generation, perhaps enough to tip the balance between a two-degree and a three-degree scenario in terms of the eventual global warming toll. This 'gigatonne gap' attributable to loss of nuclear will make our even more stringent 350 boundary unobtainable, with all the ensuing climatic consequences outlined earlier in this chapter. To make such a choice based on an irrational view of nuclear power's risks would be a devastating mistake for humanity to make. Environmentalists have been on the wrong side of this debate before, and they should think hard before allowing history to repeat itself. Any reasonable science-based assessment, such as Greens insist should guide us when considering climate change, refutes most of what the anti-nuclear lobby dishes out as 'fact'. The British environmental writer George Monbiot has even compared anti-nuclear activists to global warming deniers in terms of their treatment of the science on radiation. To get a more ecologically-relevant idea of the upsides and downsides of nuclear, we need to bear all the planetary boundaries in mind – and as later chapters will show, it scores highly even compared to renewables.

As I have insisted repeatedly, the real debate should not be between nuclear or renewables: to have a realistic chance of limiting global warming we certainly need both. Much of the world's power could also come from a comparatively small area of solar thermal plants constructed in the world's hot deserts. For Europe, a fifth of the continent's electricity could come from solar stations in the Sahara, with the power carried underneath the Mediterranean Sea via high-voltage cables. The area required would be 2,500 square kilometres of solar collecting mirrors, rising to 6,000 square kilometres if solar power were also supplied throughout the Middle East and North Africa as well as exported to Europe.[64] This sounds like a lot, but is equivalent to only the area of Lake Nasser above the Aswan Dam in Egypt, and would create thirty times as much power as that dam does using hydroelectricity. The southwestern United States also has enormous solar potential, and the world's largest solar thermal plant is located there: the Solar Energy Generating Systems plant in California's Mojave Desert. As of October 2010, the California Energy Commission reported that an additional 34 large solar thermal plants have now been proposed, with a potential generating capacity of 24 gigawatts (a large coal plant might provide 1 gigawatt).[65] However, there are land-use conflicts to be concerned with regarding this area, as Chapter 5 will show.

Perhaps the best place anywhere in the world for solar electricity is Australia, where a hot and sunny desert interior is surrounded by densely populated coastal cities. A square of outback 50 kilometres by 50 kilometres could theoretically supply all of Australia's electricity demand using concentrating solar-power mirrors.[66] In practice, many different solar thermal plants would be built inland from the major cities, reducing power transmission losses and spreading the infrastructure more widely. Solar hot-water systems are also an important technology for Australian households, although only 5 per cent use them at present. The sun's heat can also be used for cooling, thanks to a technology called absorption chilling. This makes additional sense, because the need for cooling is likely to be largest during the heat of the day, when the sun delivers the most energy. Shockingly, however, at the time of writing in April 2011 Australia has no operating commercial-scale solar thermal plants, and remains 85 per cent dependent on coal for its electricity. The country consequently has per capita carbon emissions higher even than the United States, at a shameful 20 tonnes of CO_2 per person,[67] and its government currently projects a 25 per cent *increase* in emissions by 2020.[68]

The other key renewable technology to displace fossil fuels, and the one perhaps best-loved by Greens, is wind. According to the Global Wind Energy Council, the world's installed wind-power capacity is expected to have reached 200 gigawatts by the end of 2010, with 40 gigawatts added during that year alone.[69] This is already making a significant dent in carbon dioxide emissions, especially if the alternative is coal. But the scale of the challenge remains enormous: in 2007, according to the US Energy Information Administration, global installed coal-generating capacity totalled some 1,425 gigawatts, and is projected to rise to 2,360 by 2035.[70] Wind is now a huge and growing business, with key movers in the industry like India's Suzlon and Denmark's Vestas among the big-league corporate players worldwide. In the US, General Electric is also a major wind turbine manufacturer. In the US, new wind installations fell by half in 2010 compared to the previous year, thanks to the recession and political wrangling over energy policy (or lack of it), even while the industry went from strength to strength in Europe and Asia.[71] One of the big remaining hurdles in America is a straightforward lack of wires: new wind installations often lack grid connections to take their clean electricity to markets. In October 2010 Google Foundation became a partner in the Atlantic Wind Connection, a plan for a high-voltage undersea cable to bring large-scale offshore wind to the US East Coast.[72]

In the UK onshore wind is increasingly being stymied by local opposition: according to one report, at least 230 local anti-wind farm campaigns now operate across the country, and approvals for new wind farms have fallen to an all-time low.[73] From a climate-change perspective opposing wind is just as reprehensible as opposing nuclear, and as I discuss later in this book I do not think any of the planetary boundaries offer compelling environmental reasons for completely ruling out wind – even though problems with bird and bat kills are real and severe in some cases. In any case, the UK industry is likely to largely move offshore, where a truly vast resource awaits commercial exploitation. According to a recent assessment, offshore wind and other marine renewables could deliver six times current UK electricity demand.[74] In September 2010 the world's largest offshore wind farm opened in the North Sea off Thanet in Kent, adding a further 300 MW to Britain's 5 gigawatts installed capacity of wind.[75]

But new turbine installations are expected to plummet from 2010 to 2013, and the reason has nothing to do with a shortage of wind. Instead, the problem is money. Both renewables and nuclear are highly capital-intensive, meaning that all the investment needs to be made upfront, whilst the price of fuel is very low (for uranium) or free (for wind and solar). Capital financing is therefore a huge challenge. Indeed, it is no exaggeration to say that climate change is, more than anything else, a financing issue. With post-recession commercial banks shutting down much of their lending operations, energy utilities are increasingly cash-strapped and unable to finance future investment. The implication of this may be hard for free-marketeers to swallow, but a large portion of future energy infrastructure may need to be supported and directed by the public sector.

Energy is very different from other commodities, for without energy nothing can happen anywhere in the economy. It cannot be substituted. Governments, therefore, are key players. It was a French government decision (after the 1974 oil price shock) to move comprehensively towards nuclear that makes the country an accidental climate-change champion today, whilst Britain's liberalised approach has led to a real danger of blackouts – and the missing of renewables targets – as investment has failed to materialise. Energy, unlike much of the rest of the economy, demands a degree of central planning, to get the mix of different technologies right in the grid. Wind, for instance, is highly intermittent, so cannot be relied on for baseload (always on) generation. Nuclear, on the other hand, cannot be easily ramped up and down in response to peaks in demand, so is baseload only. Gas can be easily switched on and off to balance load demand, but emits carbon. And so on. Decisions about the extent to combine the different technologies cannot simply be left to the market if the target is a carbon-neutral electricity supply.

In the Maldives, the country faces precisely these questions on a micro level. With each island operating a separate electricity grid, we cannot simply put up a lot of solar panels or wind turbines and declare the country carbon-neutral. Wind speeds are only good for part of the year, and solar requires backup on cloudy days and at night. Batteries might be an option, but for larger islands like the capital Malé the number of batteries needed will make this unviable. We could also keep running diesel generators for the time being to balance electrical loads, but this means carrying on emitting a residual amount of carbon. One option for Malé is to burn rubbish on demand, using waste-to-energy technologies as backup. Another might be to import biomass and burn that, but sustainable and reliable sources from overseas would need to be found, as the Maldives has no land to spare. My conclusion is that we need an energy and carbon neutrality 'master plan'. This cannot simply be done on the hoof, or the danger is that either nothing at all will happen, or the wrong investments will be made and a lot of money wasted.

In the Maldives as elsewhere, energy efficiency is central. It will always be much cheaper to stop wasting energy than to build energy-generating capacity to cover unnecessary use, whether that generating capacity is powered by diesel, nuclear fission or the sun. (The energy-efficiency guru Amory Lovins calls these saved units 'negawatts', in a play on megawatts.) In the Maldives, antiquated fridges and air-conditioning systems place a huge burden on electricity supplies; much of this could be saved if buildings were better constructed or retrofitted to absorb less solar heat and lose less cooling through doors and windows. In cold countries, fantastic amounts of energy are wasted through badly insulated roofs and walls, not to mention draughty doors and windows. Vehicles can also be made more efficient if governments mandate higher standards; European and Chinese cars use much less fuel than American ones because regulators have intervened in the vehicle market.

In big and heavily populated continents like Europe the task of building a carbon-neutral electricity sector is actually much easier than in small island states. Countries can share power over high-voltage international connections, making load balancing between far-removed areas possible. So when the North Sea is especially windy the UK will be able to export power to Germany and Spain, whilst during a calm spell Norwegian hydroelectric power can be used to balance demand. In effect, giant reservoirs in the Scandinavian mountains and the Swiss Alps act like enormous batteries, releasing power only when needed. Constructing this energy 'supergrid', including a large component of imported solar power from North Africa, is a central task for the European Union. As the *Economist* recently pointed out, a single market in energy makes more sense than a single market in almost anything else.[76] This is estimated to cost 1 trillion euros over the next decade, however. The only thing holding back progress is finding the cash.

Money, however, is not a limited resource in the same sense as energy. Finance we can create, if we are clever enough. Jasper Sky, a colleague at Oxford University's Environmental Change Institute, suggests creating new funds with a novel twist on the traditional tactic used by recession-hit governments of 'quantitative easing' (QE). QE normally means that a central bank buys government bonds from investors, in effect creating new cash, which is then available to banks to encourage them to lend more and thereby increase economic activity. Sky suggests that a central bank could buy specially issued bonds from a Green Investment Bank, which would then use its funds to support new clean technology deployment at a large scale, from offshore wind to nuclear to supergrids. By spending on large infrastructure projects, this Green bank would help revive the economy and create jobs, whilst at the same time putting the country on the path towards a carbon-neutral energy system. In Sky's words, 'we could potentially build our way out of the climate crisis without taking on new debt and without increasing electricity prices, whilst tackling the economic crisis at the same time'. To some this 'printing money' approach might raise the spectre of inflation, but central banks have plenty of options – like limiting the annual amount offered, altering interest rates or forcing commercial banks to hold more reserves – to make this a minor concern.

A version of this idea has been put at the international level by the World Future Council, which suggests using the International Monetary Fund's 'special drawing rights' facility to create a $100–billion-a-year Green Fund for investment in low-carbon infrastructure.[77] Much of this money could go to developing countries, particularly those in the poorest category of 'least developed'. Many of these countries would be much better off eschewing expensive fossil fuel imports in favour of using their indigenous renewables, but lack both the funding and the capacity to manage energy policy properly. In case there is a danger of some of this money getting into the wrong hands, Jasper Sky proposes a system where the projects would be backed by promissory notes issued by the IMF rather than actual cash – which would only be paid out to developers once a project was properly up and running.

NEW TECHNOLOGIES FOR THE FUTURE

Despite frequent claims to the contrary, current technology will not suffice on its own to do the job. We will need dramatic improvements in the efficiency of renewable power sources and breakthroughs also in technological options for removing the additional burden of carbon dioxide that has already accumulated in the atmosphere. This calls for big investments in research, development and demonstration (RD&D) in energy. The good news is that, after falling continuously since about 1979, energy RD&D began an upturn in 1997 that continues today – and indeed has been hugely boosted by government economic stimulus spending, now totalling $23 billion in OECD countries, according to the International Energy Agency (IEA). Less good news is that energy as a proportion of total RD&D has actually declined, from 12 per cent in 1981 to 4 per cent in 2008.[78] And with the stimulus spending quickly giving way to cuts and public-sector austerity, the outlook for government-funded energy technology innovation seems bleak.

Closing this funding gap on RD&D is crucial to meet any long-term targets to reduce carbon emissions, not just the 350 ppm planetary boundary. This means an additional spending of $40 to $90 billon, according to the IEA, but much more is needed to meet more stringent targets for CO_2 reduction. How can this money be raised? One suggestion, proposed by the Canadian economists Isabel Galiana and Chris Green, is for a low – say $5 per tonne – price on carbon, which would barely be noticed by consumers (and hence not attract political opposition) but could raise $150 billion globally per year for RD&D.[79] An analogy might be the nuclear industry, which has to pay a small charge for each megawatt-hour of nuclear-generated electricity towards eventual decommissioning costs. The airline industry, for example, could do the same, with a small fee per passenger or flight going to a technology fund that will find ways to decarbonise the industry and so reduce its damage to the atmosphere.

Galiana and Green also suggest that this carbon price should rise gradually over time, sending a forward price signal to the energy markets. Another option is to remove the trading elements from cap and trade schemes like the European Emissions Trading Scheme, shifting the system into a simple auction of carbon credits with the proceeds allocated to an independently administered energy RD&D trust fund. So far the evidence from Europe is that its flagship carbon-trading scheme has done little or nothing to reduce emissions, whilst allowing the big energy corporations to pocket billions in windfall profits because pliable politicians handed them carbon credits for free rather than auctioning them.[80] Never mind the 'polluter pays' principle: in this case the polluter got paid. This is not just iniquitous; it represents a serious opportunity cost, as the money went into the pockets of shareholders rather than being usefully invested.

So what new technologies should we focus on, and what new breakthroughs are needed? Top of my list would be figuring out a cost-effective way to store electricity at a grid level, solving the fundamental problem of intermittency from renewables like wind and solar. From the Maldives to continental Europe, the biggest headache with wind turbines and solar panels is that they deliver nothing when the wind stops blowing and the sun stops shining, even though people go on consuming electricity as normal. If some way could be found to store their output, at the very largest scale, then 100 per cent renewable power would be a realistic option, rather than backup being needed from on-demand sources like gas.

Fossil fuels have been so successful mainly because they represent concentrated stores of chemical energy, allowing power to be supplied whenever we want and transported globally via pipelines and ships. We need to find a way to allow renewables to deliver the same quality of energy reliability. Hydrogen has been proposed as one storage option, but half the energy is lost when the hydrogen is generated and then converted back again into electricity. Pumped storage reservoirs can also be used to store potential energy in water, but only work in mountainous areas and cost a lot to build. Figuring out how to store electricity reliably, efficiently and on a massive scale is probably humanity's greatest single technological challenge.

Second on my list, and related, is the development of electric vehicles that are easy and quick to recharge and can travel several hundred kilometres without having to carry most of their weight in heavy battery packs. We need much more efficient and cheaper batteries that do not use rare earth minerals like neodymium, terbium and dysprosium, currently essential components of electric cars but likely to grow scarcer in decades to come.[81] Still looking at transport, an even greater technical challenge will be to decarbonise aviation, or at least find a way to allow speedy international travel over oceans without heavy carbon emissions. As with cars and electricity, people are not going to give up flying long distances, so some solution must be found to tackle aircraft-generated carbon. One conceivable option is liquid biofuels (if truly sustainable sources on the scale needed can be found); another is hydrogen-burning jet engines. At the moment there is no clear answer, but the entirely implausible today may become obvious tomorrow: that is the job of R&D.

Third for me would be the development at large scale of carbon capture and storage (CCS) technology, to strip CO_2 out of power-station waste gases and pump it safely into saline aquifers or stable rock formations underground. There is a strong argument that countries like China that possess a large amount of coal are going to burn it anyway – so finding ways to stop the resulting carbon dioxide hitting the atmosphere has to be a big priority. Currently several test projects are up and running around the world, but despite the political attention given to 'clean coal' none are running at the scale of a large power station. An additional reason to make CCS work is so that biomass power plants could be built that would actually be carbon-negative. The vegetation burnt in them would have absorbed carbon from the atmosphere whilst the plants grew, so injecting the resulting carbon dioxide underground would strip this from the atmosphere. If deployed at a large enough scale in future decades (and assuming this did not overly affect land use and food production), this could be a useful tool for meeting the 350 ppm goal.

Fourth on my list but equally revolutionary in its potential is next-generation nuclear technology, which could improve substantially on the designs used in civil nuclear plants in the past. Particularly exciting is the Integral Fast Reactor (IFR) concept, for which a prototype was nearly completed at the USA's Argonne National Laboratory in the early 1990s but was cancelled for political reasons by the Clinton administration. Because IFRs would be 'fast breeder' reactors they could utilise much more of the potential energy in uranium fission: in conventional reactors, only 1 per cent of the fissionable energy is used from the natural uranium and the rest becomes waste. Another attractive potential for IFRs is for them to generate zero-carbon electricity by burning up stockpiles of existing long-lived nuclear waste, whilst producing waste themselves whose radioactivity declines back to that of the original uranium ore in just 200 years.[82] Another option seriously worth exploring is thorium as a reactor fuel, as it is much more plentiful than uranium-235, produces less waste than conventional reactors and does not yield anything that might be useful for nuclear weapons in the hands of rogue states.

Even if we stick with uranium, there is no conceivable shortage of fuel for IFRs. Whilst there are real fears that limits on accessible reserves of conventionally used uranium-235 might put a ceiling on worldwide nuclear reactor deployment, IFRs can burn up so-called 'depleted uranium', or uranium-238, of which there is no shortage at all. Here's a real killer fact: there is enough depleted uranium sitting around in one yard in Kentucky to run the United States at current consumption entirely on IFRs for six centuries.[83] That is without opening a single new uranium mine. IFRs seem to be to be prime candidates for urgent and substantial RD&D funding.

THE POLITICS OF CARBON

I know many environmentalists will read all this with a sense of growing horror. A technofix! How outrageous! What a cop-out! There seems to be a perverse pleasure taken by Greens in reminding us all just how difficult dealing with climate change is going to be, and how impossibly enormous the required behavioural change in the way we use energy. It is almost as if there is something shameful about proposing a solution that is too easy, and that lacks the stark moral challenge of the conventional Green narrative. But making climate mitigation difficult and unpleasant is hardly a recipe for success. I particularly dislike the high-profile switch-off campaigns where whole cities are plunged into darkness for an hour as a supposedly symbolic gesture about energy use. So is the implication that we all need to live in constant gloom to reduce CO_2 emissions? Similarly with the oft-repeated insistence that we should all turn down our thermostats. Why? Should we be cold and uncomfortable for ever?

I would instead suggest that we focus on continuing to generate the energy for people to live comfortable, prosperous lives, but simply do so in a low-carbon (and later zero-carbon) way – with the proviso that energy should not be wasted unnecessarily, and using it more efficiently can deliver many of the lifestyle benefits of producing more. I reject the implication that carbon reduction should be held hostage to a wider ideological programme seeking to change people's lifestyles and patterns of behaviour.

Similarly, if it is to succeed in helping us to meet the climate change planetary boundary, the environmental movement needs to become comfortable with centralised technologies and big corporations. Focusing on small-scale solutions may feel good for individuals, but it is not going to solve a planet-scale problem. Small may be beautiful, but big is better when you have the entire Earth's future at risk and many billions of people to provide sustainable energy for. Instead, much of the climate-change activist movement has grown more and more extremist in its rejection of corporations and markets. During the Copenhagen climate summit in December 2009 thousands of activists organised themselves at a 'Klimaforum' outside the main conference centre where national delegates were meeting. Their 'People's Declaration' insisted that 'no false, dangerous, or short-term solutions should be promoted and adopted, such as nuclear power, agro-fuels, offsetting, carbon capture and storage (CCS), biochar, geo-engineering, and carbon trading'.

Instead, 'real solutions', the activists maintained, should be 'based on safe, clean, renewable, and sustainable use of natural resources, as well as transitions to food, energy, land, and water sovereignty' (whatever that might mean).[84] 'System Change – not Climate Change' was the mantra, the system in question being, of course, the market and capitalism. 'We want to take the future into our own hands by building a strong and popular movement of youth, women, men, workers, peasants, fisher folks, indigenous peoples, people of colour, and urban and rural social groups,' the activists declared, making sure to leave nobody out. I know this is well-intentioned, but it is little more than a recipe for continued marginalisation. Moreover, hardcore ideological campaigning can spark a backlash against climate mitigation in general amongst those not amenable to this kind of politics.

None of this to argue that technology can do the job on its own. Politics needs to drive the changes in technology and investment that will solve the climate problem. The two are not opposites, as is too often asserted, but entirely complementary. In the UK for example, a strong Climate Change Act now legally mandates the government to achieve 80 per cent reductions by 2050, and to achieve several intermediate milestones along the way. Much credit for the Act goes to Friends of the Earth, who mobilised hundreds of thousands of British voters to write to their members of parliament on the subject. Adopting legally mandated climate-change targets has helped make the UK an acknowledged international leader in global-warming negotiations, and is now bringing about a belated revolution in renewables as well as a renaissance of nuclear power. Cynics often grumble: 'The government isn't doing anything', but in Britain that is simply not true.

Although the combined lobbying force of industry was once aligned against action on carbon emissions, today there are many very powerful companies that strongly and publicly support action on climate change. Google, one of the richest companies on the planet, is sponsoring an ingenious programme called 'RE<C', the acronym standing for the need to make renewable energy cheaper than coal at a grid scale. Whilst the US Congress failed to pass 'cap and trade' legislation in late 2010, dashing the hopes of many in the process, it is far from clear that nefarious big business lobbying was mainly to blame. The US Climate Action Partnership,[85] which brought together major corporations like Ford, Shell and Duke Energy, lobbied in favour of the legislation – but still lost. What went wrong was that the political polarisation of US politics infected the climate-change issue so much that not a single Republican senator dared to vote for the proposed law.

Some economists have argued for a wait-and-see approach with climate change, arguing that adopting clean technology in a few decades' time when it is cheaper is better than deploying expensive alternatives today. This argument flies in the face of basic climate science. A group of scientists led by Oxford University's Myles Allen recently launched a website to popularise their argument that cumulative carbon emissions – not levels in any specific year or decade – matter most to the climate system.[86] 'To limit long-term carbon dioxide levels to 350 parts per million,' the scientists write, 'we will need to limit cumulative carbon dioxide emissions to less than one trillion tonnes of carbon' in total throughout the whole Anthropocene.

In order to do this, the date when emissions peak becomes all-important: if global reductions start in 2015, for example, then the decarbonisation rate need only be 3 per cent per year to stay within the overall carbon budget.[87] This is quite achievable, if ambitious. But if we postpone action to reduce global emissions until 2030, the required reduction will be 8 per cent per year, unfeasibly expensive by any definition because it will mean throwing away a huge amount of energy infrastructure before it has reached the end of its lifetime. The naysaying economists have it backwards, in other words. The longer we wait to deal with carbon, the more expensive that task is going to be – if we are serious about limiting temperature rises and getting back within the planetary boundary. There is no time to lose.

CHAPTER FOUR

The Nitrogen Boundary

The story of carbon is the story of humans transcending one of the fundamental limits of the biosphere. Having previously been restricted to wood, water and wind as energy sources, we discovered fossil fuels and used them to build a complex and advanced industrial civilisation. The story of nitrogen is just as extraordinary. For all of history, this element has been the main limiting nutrient on plant growth – shortages of nitrogen in the soil meant that human agriculture would always struggle to support the population. Right up until the early twentieth century, this problem seemed insuperable. Not for nothing did Malthus predict a large-scale die-off of people in over-populated Europe. But the problem was solved through the mass production of industrial crop fertilisers, and the story of how this happened – and its implications for the planet's environment – holds lessons for us all.

Those of us who celebrate human technological ingenuity, as I do, must admit that human inventions can have dramatic and unpredictable impacts on the biosphere. Whilst our burning of fossil fuels has increased the atmosphere's concentration of carbon dioxide by 30 per cent since pre-industrial times, the volume of the terrestrial nitrogen deposited by humans has more than doubled. This has had enormous impacts on land, oceans and rivers, enough for limitations on nitrogen use to be one of the planetary boundaries recommended by the expert scientific team as necessary to protect the whole Earth system. But the lesson for environmentalists is equally profound, for the story of nitrogen shows that their favoured prescription for agriculture – a worldwide switch to organic farming – cannot possibly feed the world. We are left with a quandary: humanity is already using too much nitrogen, yet we need even more to feed a growing population. Luckily there are ways through the dilemma, but they are ones that many readers may find unpalatable.

First, a quick chemistry lesson. Like carbon, nitrogen is all around us: 78 per cent of the air we breathe is composed of N_2, an odourless, colourless and entirely uninteresting gas. In this form it does absolutely nothing that is even vaguely worth writing – or campaigning – about. Even so, nitrogen in a different form is essential to life. Every living cell needs nitrogen: it makes leaves green, constitutes an essential part of all proteins, forms enzymes and helps encode genetic information in DNA and RNA. Without nitrogen, our crops would die in the fields and our children would develop the awful starved pot-bellies of African refugee camps.

One of the few ways reactive nitrogen appears naturally is via the instantaneous high temperatures and pressures created by electrical discharges in thunderstorms. Believe it or not, lightning was our first source of fertiliser, depositing small quantities of nitrates in the soil with rain. The second was less conspicuous: it takes place quietly under the soil, in the nodules that can be found attached to the roots of leguminous plants like clover and beans. And that is about it. Apart from thunder and the actions of a few very specialised microbes, there is no other way to get nitrogen into the active biosphere. No wonder nitrogen shortage is the greatest limiting factor on natural biomass productivity both in land and at sea.

That too is why the story of human agriculture is one of repeated – even structural – famine. Throughout history famine has been a constant presence: in Europe 'great famines' saw millions starve to death in conditions of appalling suffering between 1315–17 and the 1590s. In 1600 half a million Russians died of famine-related starvation.[1] It was known that manure was a fertiliser, and in more recent times that intercropping with legumes could increase productivity: but this uses more land, for farmland sown with clover could not also yield wheat for much-needed bread. Even so, the amount of nitrogen fixed by cultivated legumes doubled from 1860 to 2000 (from 15 to 30 million tonnes) as farmers struggled to raise yields.[2]

The constant spectre of famine was not unique to Europe; it was historically present also in China and other agricultural societies the world over. Stuck with organic farming, pre-industrial humans were battling not just nature, but chemistry itself. In the millennium up to 1800, human GDP had hardly changed at all, as a result of chronically low agricultural productivity. Expressed in 1990 dollar terms, per capita GDP in western Europe was $400 in AD 1000, $676 in 1400 and $771 in 1500. By 1820, however, after a thousand years of minimal growth, per capita GDP had doubled, to a still-paltry $1,204. For comparison, consider that by 2001 the GDP of 1820 had increased sixteen-fold, to $19,256.[3] Something had happened – something that freed humanity from the iron grip of what had till then been one of the strictest laws of nature.

THE SPECTRE OF FAMINE

By the end of the nineteenth century, scientists were well aware of the problems that nitrogen limitation posed for humanity. 'England and all civilised nations stand in deadly peril of not having enough to eat,' warned William Crookes, a scientist and president of the British Association, in a widely quoted 1898 speech. At issue were predicted declines in harvests following the likely exhaustion of imported Peruvian guano stocks and nitrates stripped from Chilean deserts. (Guano is an excellent fertiliser, being composed of seabird excrement – often tens of metres thick – deposited over many centuries on arid offshore islands.) Human populations were rising rapidly, and fears that human fertility could outpace soil fertility were very real. 'As mouths multiply, food resources dwindle,' Crookes went on. 'I am constrained to show that our wheat-producing soil is totally unequal to the strain put upon it.'

But Crookes was not Malthus. He knew that there was a way out of what he called this 'colossal dilemma'. And that path to plenty lay through science. 'It is the chemist who must come to the rescue of the threatened communities,' Crookes insisted. 'It is through the laboratory that starvation may ultimately be turned into plenty.'[4] If William Crookes's optimism seems old-fashioned, I suspect this is because if a scientist of his calibre were to make a speech in an area of equivalent concern today, he would likely be shouted down as a techno-fantasist. But the historical lesson is that Crookes was right. His challenge may have seemed impossible at the time, yet it was solved by the application of human ingenuity, and within just ten years after he spoke.

However, the chemists who performed this feat, and were later awarded the Nobel Prize in recognition of their stunning breakthrough, worked not for the King but for the Kaiser. And their work was initially put to the service not of feeding the world, but of producing industrial-scale instruments for mass slaughter in the First World War.

ENTER FRITZ HABER

Fritz Haber and Carl Bosch can claim as dramatic an influence on twentieth-century world history as the likes of Stalin, Churchill and Gandhi, yet their names are barely remembered, and their defining achievement little celebrated by today's generations – even though we benefit from their inventiveness every time we take a meal. The chemical technique that bears their names, the Haber-Bosch process, was undoubtedly 'the most important technical invention of the twentieth century', according to the Canadian scholar Vaclav Smil, whose book *Enriching the Earth* is the definitive biography of nitrogen in the modern world. As Smil relates, the 'single most important change affecting the world's population – its expansion from 1.6 billion people in 1900 to today's 6 billion – would not have been possible without the synthesis of ammonia' for conversion into artificial fertilisers.[5]

Haber was not the first chemist by any means to take up Crooke's challenge. But he was the first to refuse to be defeated by it. The problem should have been easy to solve: all the essential chemistry demanded was for N_2 to be combined with hydrogen to form ammonia, or NH_3. But the twin bonds between nitrogen atoms in atmospheric dinitrogen are extremely strong, and the high temperatures and pressures needed to break them so extreme as to make the process appear technically unfeasible. In 1904 Haber, a professional chemist from a reasonably well-to-do German–Jewish family in Breslau, began work on the problem of ammonia synthesis. Without an offer of financial support from the German chemicals company Badische Anilin- & Soda-Fabrik (better known by its initials, BASF) to the tune of 6,000 marks a year, Fritz Haber too might have been forced to abandon the quest.

The day the world's nitrogen cycle changed can be dated very precisely: 2 July 1909, when Fritz Haber – in front of two sceptical BASF representatives – demonstrated the synthesis of ammonia in a pressurised metal tube just 74 cm tall and 13 cm in diameter. His breakthrough had been to discover a catalyst (originally uranium, later switched to a form of iron) to assist the reaction: even so, the tube had to be heated to 500°C and the pressure pumped up to the equivalent of 100 atmospheres for it to work. A mere 2 cubic centimetres of ammonia per minute trickled out at the end, hardly enough to fertilise a window-box, let alone all the depleted fields of Europe. But the process worked. Now Carl Bosch, a canny executive working for BASF, was to step in. His job was to turn Haber's discovery into something that would work at a truly industrial scale.

Success would require the application of strategic technical, organisational and engineering genius on the part of Bosch that was at least the equal of Haber's application of pure chemistry. Fortunately, BASF saw the potential of the new invention and put substantial resources at Bosch's disposal. Within just four years he had ironed out serious problems – from exploding pressure vessels to the need to find a cheaper metal catalyst – and built a large-scale plant that, by October 1913, was producing 10 tonnes of ammonia per day. Bosch was undoubtedly motivated by the profit-making potential of his new and quickly patented process. But like Haber, he also saw its application as being for the good of humanity. As Haber said later in his Nobel lecture,[6] his concern was to avoid 'a major emergency' in food supplies when the limited 'supply of saltpetre nitrogen which Nature had deposited in the high-mountain deserts of Chile' began to run short. Now, he hoped that 'improved nitrogen fertilization of the soil brings new nutritive riches to mankind and that the chemical industry comes to the aid of the farmer who, in the good earth, changes stones into bread'.

But the road to changing stones into bread was to be a rocky one, for clouds of war were gathering over Europe. In Germany, where Bosch was working, the Kaiser's generals knew that an Allied naval blockade could easily deprive them of the imported Chilean nitrates that were essential for producing munitions. Bosch's synthetic ammonia was to be expressly dual-purpose: converted into ammonium sulphate it could fertilise German crops, but turned instead into ammonium nitrate it could also manufacture German bombs. Haber himself was one of the signatories to a patriotic declaration by German scientists that declared 'the German army and the German people are one'.[7] This patriotism would drive him into making some very damaging moral decisions over the years to come, ones that blacken his name to this day.

Shortly after the outbreak of war in 1914 the feared Allied naval blockade materialised, and its effects were just as devastating as the German generals feared. With their oceanic supply lines to Chile cut off, the Kaiser would have run critically short of munitions as early as the spring of 1915. Instead, BASF and the Second Reich began a strong partnership. It was Bosch himself who suggested, following a French air raid on his west German plant, building a new, much larger-scale industrial operation deeper inside Germany. The War Ministry put up 12 million marks, and the new plant at Leuna began operation on 27 April 1917, producing more than 100,000 tonnes of nitrates annually.[8] Thanks to the Haber-Bosch process, the carnage of the First World War was to continue for another year and a half before the German capitulation finally came.

But Fritz Haber's personal responsibility for the atrocities of the First World War was unfortunately much greater than just his indirect contribution to securing German explosives supplies. In his patriotic fervour, Haber became head of Germany's Chemical Warfare Service, turning his agile mind to the task of poisoning Allied troops with gas attacks. The first large-scale use of Haber's chlorine gas, at 5 pm on 22 April 1915 on the Western Front at Ypres, nearly broke the French lines: as the curious yellow-green cloud engulfed them, panic-stricken French and Algerian troops – coughing up blood and choking to death in their hundreds – fled in disarray. Had the Germans been ready to follow up their success, the tide of war might have turned there and then.

With Germany facing outrage from half the world, Fritz Haber was even entreated by his own wife 'to forsake poison gas'.[9] But Haber would not be budged, and on 2 May 1915 a stricken Clara Haber committed suicide, using her husband's own military-issue revolver. Following this personal tragedy, Haber travelled to the Eastern Front to supervise gas attacks on Russian soldiers there. (It should be remembered that, at the same time as condemning Germany for its barbarism, Britain quickly developed and used its own poison gas weaponry.) His reputation was ruined, and Haber died in 1934 following years of depression and increasing Nazi persecution because of his Jewish origins. His postwar award of the 1918 Nobel Prize for Chemistry was so controversial that the ceremony had to be postponed until June 1920 because of protests from the former Allied countries.

THE NITROGEN APE

Thanks to Fritz Haber, humans are the only species on the planet, apart from *Rhizobium* bacteria, that are able to fix their own nitrogen directly from the atmosphere. (*Rhizobium* are the symbiotic inhabitants of the root nodules found in legumes.) In that sense we are perhaps biologically closer to peas, beans and clover than to our great ape relatives. Certainly everywhere humans dwell, the Earth is richly fertilised: in general it is a surfeit rather than a lack of nutrients that lies behind many of our most intractable environmental problems. The production of synthetic fertiliser now mobilises some 100 million tonnes of nitrogen every year, and as a direct result the carrying capacity of the world's croplands rose from 1.9 to 4.8 persons per hectare in the century from 1908 to 2008. Turning stones into bread, in Haber's phrase, is now so commonplace that we give it scarcely a moment's thought.

The process was not one of steady expansion after Haber's breakthrough. Ammonia production stagnated in the 1920s, and global output really only began to take off after the Second World War, led by the United States. In the 1960s and 1970s, the Green Revolution then brought high-yielding hybrid crop varieties to developing countries that were enormously productive when grown with artificial fertilisers. China, faced with a growing population and no free land, built thirteen ammonia plants in the 1970s, and by 1989 had become the largest nitrogen producer in the world. The United States remains heavily dependent on agricultural fertilisers – so much so that major crop-growing regions are actually served by ammonia pipelines. Today, more than half the nitrogen produced by the world's crops originates in ammonia production plants using the Haber-Bosch process. Given that most of this ends up as our food, it is fair to say that most of the protein in the modern human being's body is of synthetic origin.[10]

Even whilst the production of artificial fertilisers has clearly been good for humanity, it has also begun to cause serious environmental problems. Because reactive nitrogen is so rare in nature, ecosystems are highly sensitive to it, whether on land or in the water. Nitrogenous fertilisers quickly wash off fields and into watercourses, where they stimulate blooms of algae and deplete the water column of oxygen. This 'eutrophication' process has dramatically affected rivers and lakes across the Northern Hemisphere, and where this nutrient-rich soup pours out into coastal oceans, enormous oxygen-depleted 'dead zones' mount a silent takeover of continental shelves. The latest scientific count found a worrying 400 separate dead zones spreading out from the world's most densely populated coastal regions, from Shanghai to the mouth of the Mississippi. The Gulf of Mexico dead zone now affects an average of 20,000 square kilometres each summer. Large areas of the Baltic Sea in northern Europe are similarly affected by runoff and nutrient pollution.

In the latter, as in many inland lakes, phosphorus pollution – from detergents as well as fertilisers – is arguably an even bigger problem. In the oceans the only species that can survive a seasonal dead zone are those that can crawl, burrow or swim away as the poisonous waters creep invisibly upwards towards the sea surface.[11] The global dead zone total now covers more than 245,000 square kilometres, an area the same size as the United Kingdom.[12]

Reactive nitrogen is extremely potent stuff, and can keep on polluting over and over again as it transforms and circulates between different compounds. A single liberated nitrogen atom may, for example, join with oxygen to form nitrogen oxides (the precursors of toxic low-level ozone smog), before later raining out of the sky as nitric acid, going on to fertilise grasslands and watercourses as nitrates, and then being converted into the powerful greenhouse gas nitrous oxide. Even nitrogen that is sensibly applied to fields will eventually form part of this cycle: after being incorporated in a crop it will inevitably one day be released again – whether into the sewage system as waste or by a decomposing corpse in a grave. The only way this apparently immortal nitrogen atom can ever be properly silenced is for it to be combined again with another nitrogen atom back into the happy pair bond N_2. Then it can once again enter the atmosphere as an inert and entirely trouble-free gas. But humans cannot perform this 'denitrification' trick – only certain microbes can do it, and then only under very specific circumstances such as in the absence of oxygen.

It is not all bad news, however. The human production of vast quantities of nitrogen is greening the Earth, and may be helping plants to draw down increased volumes of carbon dioxide – thereby offsetting climate change. One group of scientists suggests that nitrogen deposition by humans (not directly as fertiliser, but indirectly through stray nitrogen that blows, washes or rains into natural ecosystems) is having just this effect in forests, with two hundred kilos of carbon mopped up for each additional kg of accidentally deposited nitrogen.[13] In total, this free benefit may be responsible for cleaning up 0.7 billion tonnes of carbon a year via Northern Hemisphere forests and woodlands – about a tenth of the total emitted.[14] If you're wondering how this nitrogen gets into the forests, the answer is via the rain: ten or more kilograms fall annually per hectare in industrialised countries, an order of magnitude more than would have been deposited by lightning in pre-industrial times.[15]

And if you don't mind the dead zones, there are benefits too to having nitrogen in the oceans. With an estimated third of oceanic nitrogen overall of human origin, the extra plankton growth thereby stimulated may be mopping up an additional 300 million tonnes of carbon per year.[16] In areas where the seas are chronically short of nutrients, human-derived nitrogen washed down by rivers may even support whole new marine ecosystems and associated fisheries. Around the Nile Delta, for example, fisheries collapsed after the construction of the Aswan Dam in 1965 caused fewer nutrients to be washed downstream. Luckily, human pollution came to the rescue. Thanks to all the fertiliser runoff and sewage sludge now coursing down the Nile, fisheries rebounded: one estimate is that at least 80 per cent of all fish landed in the area are now supported by this anthropogenic excess nitrogen.[17]

But let's not get carried away. For the most part, nutrient pollution is bad – even when it has a fertilising effect. Study after study[18] has shown that on land enhanced nitrogen deposition tends to lead to a loss of biodiversity as weedy, fast-growing species become dominant and crowd out more specialised plants.[19] In the UK, the best-restored wildflower meadows are those that have had their fertility artificially lowered, not enhanced. Where too much human nitrogen now arrives with each successive rain shower or flood inundation, ecosystems experience 'slow but chronic loss of biological diversity', in the words of one recent survey.[20] The impacts in the oceans are likely to be similar – nitrogen and phosphorus runoff have been associated with the 'rise of slime' in the marine realm, as algae and toxic bacteria proliferate in the nutrient soup.

The production and use of nitrates also worsens climate change, both directly and indirectly. The Haber-Bosch process is energy-intensive, and gobbles up 5 per cent of the world's entire annual natural gas production – emitting millions of tonnes of additional CO_2 as a by-product. One of the consequent forms of reactive nitrogen, nitrous oxide (N_2O), is an extremely powerful greenhouse gas – fully three hundred times as potent in trapping the sun's heat as carbon dioxide – and hangs around in the atmosphere for on average more than a century. As with CO_2, methane and other greenhouse gases, concentrations of nitrous oxide have been rising, from 270 ppb (parts per billion) in pre-industrial times to 319 ppb in 2005. Agriculture is one of the biggest contributors to nitrous oxide emissions, and it barely matters whether farms use artificial fertilisers or cattle manure.[21] Nitrogen is nitrogen, pure and simple.

With all these impacts in mind, the planetary boundaries expert group proposes that the flow of human-fixed nitrogen should be reduced to slightly more than a third of its current value – from 100 million tonnes to about 35 million tonnes per year (phosphorus is included with a separately defined boundary). In case this seems a little rash, and given the many complexities and the heavy dependence of humanity on nitrogen use, the experts include the proviso that 'much more research and synthesis of information is required to determine a more informed boundary' in due course.

This nitrogen boundary affects many others, as the previous discussion illustrates – consider the impacts on biodiversity, the ozone layer, climate change and freshwater, just for starters. Consider also that by mid-century there will likely be another 3 billion mouths to feed, not to mention the higher production needs implied by changing diets and the worldwide increase in biofuels. At a first approximation, the nitrogen boundary could be even more difficult to meet than the carbon one: after all, we are only technically dependent on carbon, but we are biologically dependent on nitrogen. There are no possible substitutes for the nitrogen we need to build muscle proteins, secrete enzymes and construct cellular DNA. This may not be a boundary that we can ever meet in practice with the number of people alive on the planet today. But there is a lot we can do to try.

MEETING THE NITROGEN BOUNDARY

There are three ways that humanity can move away from our profligate use of nitrogen and towards a safer level as defined by the planetary boundaries expert group. We can synthesise less ammonia and produce less reactive nitrogen overall. We can create the conditions for microbial denitrification on a wider scale, to get reactive nitrogen turned back into harmless N_2 faster and across larger areas. And we can also use what is produced more efficiently, reducing runoff and ensuring that more nitrogen is incorporated into the crop rather than wasted. There are many opportunities for all of these, some of which confirm conventional Green and development wisdom – but some of which environmentalists will find very hard to swallow indeed. I am not doctrinaire about any of these options, but I would ask readers to consider what follows with an open mind.

One of the easiest win–wins would be eliminating the nitrogen that is produced unnecessarily in various processes, brings no benefits at all to agriculture, and is also dangerous to human health. A significant proportion of humanity's additional reactive nitrogen is created as a by-product of burning fossil fuels, such as the harmful nitrogen oxides (NOx) that come out of your car exhaust pipe. Technologies already exist that could slash the nitrogen produced from this source by two thirds, reducing the smog pollution that is highly damaging to human lungs and hearts.[22] For power stations burning coal, 'selective catalytic reduction' systems pump ammonia through the boiler exhaust gases and turn the nitrous oxides back into harmless N_2 or water. These should be made mandatory everywhere. Petrol cars in most Western countries already use smaller catalytic converters to reduce harmful pollutants, including NOx. Diesel cars, which do not usually come fitted with catalytic converters, produce many times more nitrogen pollution but a fifth less CO_2, facing drivers with a difficult choice about which is 'greenest'. Ultimately, as with power stations, the only long-term solution will be to eschew fossil fuels altogether and switch to electric vehicles.

On the international stage there are already some relatively successful efforts to control nitrogen pollution: the 1999 Gothenburg Protocol, part of the 1979 Convention on Long-range Transboundary Air Pollution, focuses on cutting both sulphur and nitrogen emissions. The Convention was the result of concerns about acid rain, to which both chemical elements are contributors. Although hardly noticed by anyone (or perhaps because), the measures have been quite a success: between 1990 and 2004 nitrogen oxide emissions fell by a third, whilst ammonia emissions have dropped by a fifth.[23] Overall, 25 million tonnes of reactive nitrogen are still produced by fossil-fuel burning, however.[24] All of this can be eliminated as a by-product of meeting the climate change boundary – a happy synergy that we should exploit to the full.

Another promising strategy is to create the conditions for microbial denitrification on the largest scale possible. Streams and rivers are vital nitrogen sinks: wetlands and flood meadows create the low-oxygen conditions in their sediments that denitrifying bacteria need, whilst also helping support freshwater biodiversity.[25] This is a win–win for at least four planetary boundaries: those on freshwater, biodiversity, land use and nitrogen. Towns and cities could pay to protect wetlands downstream from their wastewater outlets, whilst major agricultural areas – with a small levy payable by farmers on tonnages of fertiliser used – could do likewise.

Human encroachment onto fragile wetland areas at river outlets should be strictly controlled, for big deltas provide our last chance to stop nitrates washing out of rivers and causing oceanic dead zones. At the mouth of the Mississippi large areas of wetland are being lost annually because the artificially channelled river no longer delivers sediment downstream, and nitrates leached out of the heavily fertilised corn belt wash straight into the Gulf of Mexico rather than being trapped in reed beds. Reversing this process would be economically as well as environmentally beneficial because of the benefit to fisheries of reducing the size of the dead zone. None of this requires the great expense of high technology: we just need to provide the space for nature to work its magic.

Removing nitrates from wastewater is one of the major functions of sewage treatment plants, which are designed to encourage denitrifying bacteria in settling sewage sludge. However, 3.2 billion city-dwellers around the world are still not connected to modern sewage systems. Instead, in many urban slums in Africa, Asia and Latin America children play in fetid swamps whilst open sewers run past the front doors of their shacks. Millions of young lives are lost to preventable diseases as a result. Helping meet the nitrogen planetary boundary is probably the least compelling reason for tackling this injustice, but nevertheless one estimate suggests that delivering modern sewage facilities to everyone in poor countries could reduce reactive nitrogen pollution by 5 million tonnes a year.[26] World Health Organisation estimates suggest a cost of about $11 billion a year to provide both clean water and adequate sanitation to half the world's population by 2015, as demanded by the UN's Millennium Development Goals.[27] Because of the lives saved, the reduction of healthcare costs for preventable diseases, and the waste of productive time involved in fetching and carrying water, this expenditure would be strongly cost-positive. Each dollar invested would yield a return of between $5 and $46, depending on the region. There can't be many better deals than this.

It is also clear that some parts of the world use far too much fertiliser, and could reduce this without any drop in productivity. In China many growers use double or more the necessary fertiliser on their rice, cotton or vegetable crops, causing chronic nitrogen pollution problems as a result. Studies show that parts of China are so overloaded with fertiliser that inputs could be cut in half with no reduction in crop yields.[28] It is also likely that farmers in both Europe and the US could use less fertiliser, though to a lesser extent. On the other hand, growers in Africa face a serious shortage of fertiliser. If African farm productivity is to be raised, and poverty and malnutrition reduced, then substantially more reactive nitrogen needs to be supplied to replenish soils degraded by decades of overuse. Africa missed out on the Green Revolution of the 1970s and 1980s, when yields tripled in the developing world as a whole thanks to high-yielding crop varieties and large quantities of additional fertiliser.[29] Soil health will not be restored by nitrate chemicals on their own; animal manures and crop residues also need to be dug back into the soil rather than being removed for cooking fuel. Farmers also need access to markets and to credit – bringing the Green Revolution to Africa will involve a whole host of intertwined anti-poverty strategies. But more fertiliser is an essential part of the mix, so globally the potential to reduce the amount of nitrogen used in agriculture by humanity is limited.

I realise that this runs counter to the accepted Green wisdom that organic farming – which eschews artificial fertilisers – is always best for the environment. Certainly organic agriculture, if practised globally, would dramatically reduce nitrogen pollution from currently cultivated areas. But it would also leave millions starving, thanks to a major reduction in the food produced per unit of land. This conclusion emerges logically from looking at nutrient flows: with less nitrogen added, crops produce less protein in their seeds and less biomass overall, so yields go down. African agricultural productivity is very low precisely because it is stuck in organic subsistence farming and lacks additional nitrogen. Certainly this nutrient deficiency can be reduced by assiduous recycling, from both human and animal wastes. Until recently, China had a thriving 'nightsoil' industry, which collected sewage and spread it straight back onto the fields. Sewage sludge is also spread on farmers' fields today as a matter of course. But none of this adds nitrogen overall – it simply recycles it more efficiently.

A look back at history confirms this picture. In 1900 world agriculture was entirely organic and fertiliser-free, and could sustain 1.6 billion people on about 850 million hectares of cultivated land. According to the Canadian scholar Vaclav Smil, the same agronomic practices extended to today's 1.5 billion hectares of ploughed farmland would feed only 2.9 billion people. The numbers could be improved somewhat if people in rich countries agreed to surrender some of their overconsumption to poorer countries, and everyone ate less meat and dairy. (Feeding grain to livestock and then consuming meat and milk is less efficient than just eating the grain directly.) But according to Smil the basic conclusion is stark: 'Only half as many people as are alive today could be supplied by prefertiliser agriculture with very basic, overwhelmingly vegetarian, diets.'[30] That is not an appealing prospect.

This is not to say that organic farming is bad, wrong or necessarily backward-looking. In better-off countries with no immediate scarcity of land, it may be an important way to meet ecological objectives, and to provide choice for consumers who prefer not to have any chemicals used whilst growing their food. But the key reason why free-choice organic agriculture (as opposed to organic agriculture practised unwillingly by subsistence farmers who cannot afford chemical inputs) will always be a marginal activity globally is because it uses much more land to feed the same number of people. In practice this tradeoff is fatal: land use counts as a planetary boundary, as the next chapter will show, because the health of terrestrial ecosystems depends on only a small proportion of land being devoted exclusively to sustaining humans. If we were to try to use organic agriculture to feed the entire world's population at current levels of nutrition, remaining terrestrial natural ecosystems would be devastated by the massive required expansion in ploughed farmland.

This conundrum has been demonstrated in microcosm by studies of British organic farming. One piece of field research, conducted by ecologists at Leeds and York universities, demonstrated that although butterfly populations were higher on organic farms thanks to more wildlife-friendly farming practices, crop yields were typically much lower.[31] In practice, the researchers concluded, it might be better to farm conventionally (using chemicals) on smaller parcels of land, produce the same quantity of food overall, and devote the spare farmland to nature reserves – which would be much better for butterflies and other wildlife even than organic farms overall. 'If "sharing" our farmland with wildlife means that more total land will be taken into production to produce our food, then there may be a hidden cost of hurting wildlife somewhere else,' says author Professor Bill Kunin of Leeds University.[32] A second study, also published in 2010, confirmed that whilst organic farms tended to harbour 12 per cent more wildlife on average, they produced less than half the yield.[33] 'Our results show that to produce the same amount of food in the UK using organic rather than conventional means, we'd need to use twice the amount of land for agriculture,' says Professor Tim Benton, also at Leeds University.[34] This tradeoff will also apply internationally: British consumers may already be compensating for underproductive domestic organic farms by buying crops grown on land abroad, leading to a more damaging effect on biodiversity overall as rainforests and grasslands are ploughed up overseas.

Nor is organic farming necessarily better for the climate, despite the dependence of intensive industrialised-country farms on oil-guzzling heavy machinery. Researchers based at Stanford University in the US recently calculated that the widespread use of mechanisation and artificial fertilisers between 1961 and 2005 avoided the emissions of 161 billion tonnes of carbon that would otherwise have been released through land-use change.[35] (Yes, this is a *net* figure: carbon dioxide exhaust from all the tractors has been subtracted, as have agricultural nitrous oxide and methane emissions.) In other words, had farms been run organically during the period, huge areas of forest and savannah would have been brought into production to keep people from starving, destroying and releasing the stocks of carbon stored in trees and soils. During the 1961–2005 period global cropland area did increase by 27 per cent – but total yields rose 135 per cent, a jump in productivity that spared vast areas of natural land from the curse of the plough. The amount of additional land that would have been needed to produce this quantity of food at 1961 levels of yields and inputs is truly vast: 1,761 million hectares, an area larger than Russia.

Given that humanity cannot do without nitrogen fertiliser for our crops without running the risk of starvation, the most sensible strategy to try and meet the nitrogen boundary is to use what we need as efficiently as possible. That means aiming to design crops that are much more efficient in their uptake of nitrogen added by farmers to the soil. At the moment only a third of nitrogen applied to fields actually ends up in the grain. Given that cereal crops are dosed with two thirds of the global total of 128 million tonnes of fertiliser, that is an awful lot of nitrogen left to wash down rivers or add to the growing stock of nitrous oxide in the atmosphere.[36] There are ways that farmers can waste less fertiliser, but the key long-term strategy will be to develop crops with what plant biologists call better 'nitrogen-uptake efficiency'. This will be particularly important for the developing world, where the cost of fertilisers is often prohibitive for subsistence farmers, condemning them to low yields and perpetual poverty.

As things stand, there are two ways to get more nitrogen-efficient crops. One is conventional selective breeding, but this takes a long time and is rather hit-or-miss. Plant scientists have either to wait for a useful gene to spontaneously appear via some lucky mutation, or to spend years breeding generation after generation of gradual improvements into a target crop whilst hoping for an outcome that is worth the wait. With genetic engineering, however, scientists can make much more precise and rapid changes by selecting a gene from any species and inserting it into the genome of the target crop to deliver a more nitrogen-efficient and higher-yielding crop. For this reason genetic engineering is an enormously powerful technology, holding major potential to help transform humanity's ability to produce food sustainably on a limited area of land and with much less fertiliser than is used today. That it has not delivered these benefits so far is not down to any lack of effort from plant scientists, but because genetic engineering per se is implacably opposed by almost all Green groups worldwide, for ideological rather than scientific reasons.

Consider, for example, a recent breakthrough achieved in Canada by plant scientists working with canola (known in Europe as oilseed rape), the country's most important export crop. By introducing a gene from barley into canola, the scientists made the new GE crop 40 per cent more efficient in using nitrogen. In other words, it could produce the same yields as now using 40 per cent less fertiliser – a very real environmental benefit from a crop that is grown over 11 million acres of western Canada.[37] Similar nitrogen efficiency achievements have been made in plants varying from poplar trees to tobacco.[38] Perhaps the most important recent success involves transgenic rice, where the same barley gene seems to work just as well as it does in canola, delivering vast improvements in nitrogen-uptake efficiency.[39] That's something that could never be done with conventional breeding of course – you can't cross barley with rice under normal circumstances. Rice is a staple food for half the world's population, so a genetic modification that allows transgenic rice to increase its yields by a fifth with no increase in applied fertiliser could be good news both for human nutrition and the nitrogen planetary boundary.[40]

None of these potential benefits of genetic engineering cut any ice with Greens, however. 'The introduction of genetically engineered organisms into the complex ecosystems of our environment is a dangerous global experiment with nature and evolution,' argues Greenpeace.[41] Friends of the Earth has campaigned determinedly against genetic engineering for more than a decade, whilst the idea of an organic-labelled GE crop is inconceivable. I am personally very familiar with all the arguments against genetic engineering, because I used to make them myself. In an article for the *Guardian* in 2008 I wrote that 'the technology moves entirely in the wrong direction, intensifying human technological manipulation of nature when we should be aiming at a more holistic ecological approach' and that GE 'raises a whole new category of risk'.[42] A decade earlier, I joined many campaigners in taking direct action against genetically engineered crops in the UK, often at night and risking arrest by the police. Looking back, I now realise that I was caught up in more of an outbreak of mass hysteria than anything resembling a rational response to a new technology.

So what has changed? Reading the comments underneath my 2008 *Guardian* article was something of a eureka moment. One after another, the commenters pointed out that my approach was unsupported by science and largely founded in ignorance about genetics in general. That was hard to take on board, but the truth is they were right. One person wrote: 'It would be interesting to develop nitrogen fixing versions of the main food crops, thus avoiding the need for nitrate fertilisers'. That too got me thinking. Several people also posted comments pointing out that none of my arguments – such as my concern about how biotech corporations were using the new GE crops to boost their profits – logically led to an outright rejection of the technology. 'One does not fight the corporate misdeeds of the automotive industry, for instance, by demanding that the wheel must be banned,' someone pointed out in a thought-provoking analogy. I decided to stop writing about genetic engineering for a while and read up on the science before tackling the subject again.

A second eureka moment came in a book by an environmentalist who had already changed his mind about genetic engineering, the American writer Stewart Brand. In the opening sentence of the 'Green Genes' chapter in his 2010 book *Whole Earth Discipline*, Brand writes: 'I daresay the environmental movement has done more harm with its opposition to genetic engineering than any other thing we've been wrong about. We've starved people, hindered science, hurt the natural environment, and denied our own practitioners a crucial tool.' That is a strong statement, but the more I looked at the evidence, the harder I found it to disagree with him. I hunted through the scientific literature, but I could not find any convincing evidence that genetically engineered crops or foods had ever harmed a single human being or animal. Nor was there any substantiated evidence of environmental damage, even after thousands of separate tests and wide-scale commercial deployment across the world. Most convincingly of all, as Brand points out, millions of North Americans have eaten GE soya, maize and canola in huge quantities for over a decade, whilst consumers in Europe have boycotted them. Were Canadians and Americans getting sick or suffering allergic reactions as a result, we would certainly know about it by now. But they haven't been. As Brand puts it: 'It was great civilisational-scale science, and the result is now in, a conclusive existence proof. No difference can be detected between the test and the control group.'[43]

Most shocking for me was the realisation that I had been just as guilty of misusing biological science in the service of ideological ends as global-warming 'sceptics' have been in misusing climate science. I was repeating assertions made by campaigning groups without checking the primary evidence. I was citing as fact one or two studies showing apparent harm from GE products without putting these in the context of an overwhelming expert literature showing that GE was safe. (When sceptics do this on climate, I condemn it as 'cherry-picking'.) Most damning of all, I realised that throughout the entire time I had been an anti-GE activist, donning biohazard suits and mounting night-time raids against test sites, I had never read a single scientific paper on the subject. Nor did I understand much about the genome, the process of genetic engineering itself, or why transferring genes from unrelated organisms might not be so scary after all. In sum, I was not applying the high standards I was demanding from others on climate science to myself on biotechnology.

Although none of the major environmental groups will admit it still, the first generation of GE crops has almost certainly been beneficial both to the environment and to farmers. Herbicide-tolerant crops, like Monsanto's 'Roundup Ready' soya, canola and sugar beet, seem bad at first pass because they involve farmers spraying whole fields with the potent weedkiller glyphosate, which the transgenic crops are engineered to survive. The environmental benefit comes because glyphosate is pretty benign compared to most of the conventional pesticides used to kill weeds, and is so effective combined with GE crops that much less of it needs to be used.[44] This reduces toxic runoff and the number of tractor movements in a spray regime, benefiting both biodiversity and the climate. This is not to say that chemical-dependent farming using GE crops is an environmental boon, merely that it is better than the conventional non-GE alternative, as several recent studies have confirmed.[45,46] A global survey found that, between 1996 (when GE crops first began to take off) and 2005, 7 per cent less pesticide was used globally as a result, translating into a reduction of 224 million kilograms fewer toxins sprayed onto fields.[47] Given that toxic pollution is a proposed planetary boundary, this cannot be dismissed as a negligible benefit.

Another substantial benefit of herbicide-tolerant GE crops is that they allow the adoption of no-till agriculture. Instead of ploughing up farmland after each harvest, farmers can spray glyphosate to kill the remaining weeds and plant directly into the undisturbed soil the following spring. No-till helps soils conserve carbon, reducing greenhouse emissions, and also stops winter rains exacerbating erosion in exposed, naked soils. The carbon savings may be substantial: a recent review suggested about 9 million tonnes of carbon dioxide emissions are saved through soil conservation and fuel-use reduction, equivalent to removing 4 million cars from the road.[48]

Another application of genetic engineering technology has allowed farmers to cease spraying altogether, by incorporating pesticide toxins into the tissues of the crop plant itself: examples include insect-resistant maize and cotton, now planted across the world from the US to China. These have uniformly reduced the need for pesticide spraying, with benefits both to the local environment and to the human workers who would otherwise be regularly exposed to toxic sprays. In fact, the insect-resistant crops have been so successful that other pests have taken advantage of the fact that insecticide sprays are no longer being used.[49] In the context of this book, this is evidence of genetic engineering being useful to both toxics and biodiversity planetary boundaries.

One fear that many campaigners shared was that commercial GE seeds, which tend to be patented and cannot legally be saved by farmers between crops, would bankrupt poor-country growers. At activist training camps we fondly imagined a bucolic idyll of Indian and

African subsistence farmers, saving their own seed and valiantly resisting the faceless corporate onslaught forced on them by rich countries. In actual fact, farmers in developing countries were some of the first to adopt GE crops – for the practical reason that they increased yields and reduced costs. When Monsanto's GE soya was banned in Brazil under pressure from environmentalists, farmers smuggled it in over the border from Argentina in enormous quantities. The Brazilian government eventually rescinded the law.

In India the tradition of developing non-patent generic drugs was applied stealthily to GE cotton brought to the country by Monsanto. In his book *Hybrid: the history and science of plant breeding*, Noel Kingsbury writes that even whilst Western-inspired anti-GE activists were denouncing Monsanto's Bt insect-resistant cotton as 'seeds of suicide, seeds of slavery, seeds of despair', farmers were desperate to get hold of them because of their ability to kill pests without the need for expensive pesticides. 'By 2005, it was estimated that 2.5 million hectares were under "unofficial" Bt cotton, twice the acreage as under the ones which had been sown from Monsanto's packets,' Kingsbury relates. 'The unofficial Bt cotton varieties had been bred, either by companies operating in an ambiguous legal position, or by farmers themselves.' By engaging in this 'anarcho-capitalism', Indian farmers were demonstrating that 'they were not passive recipients of either technology or propaganda, but taking an active role in shaping their lives'.[50]

The idea that there is something inherent in genetic engineering technology that makes it beneficial only to big corporations is illogical but very persistent amongst environmental campaigners, many of whom are extremely suspicious of big business in principle. I shared this view for a long time, once helping co-found an organisation called Corporate Watch. But being against GE per se because it has been promoted by big companies is a bit like being against word processing because much of the most useful software is produced by Microsoft – irrational and self-defeating. In both the Brazilian and Indian cases Monsanto actually lost out, because entrepreneurial farmers without spare capital wanted its product but not to respect its patent. Farm-level benefits of GE crops, that is profits accruing not to the corporations but to the farmers, have been estimated at $5 billion for 2005 alone, and $27 billion over the past ten years.[51]

Moreover, it is not only big agro-chemical companies that are exploiting the technology. All over Africa and Asia there are publicly funded efforts aiming to create transgenic varieties of subsistence crops that will be of benefit to poorer farmers and will be made available without licensing restrictions. A quick survey of some recent initiatives includes salt-tolerant rice for use in degraded land where salinisation is reducing yields;[52] a pest-resistant eggplant developed by a university in the Philippines; Vitamin A-fortified mustard in India that could avoid 46,000 unnecessary deaths per year;[53] disease-resistant rice in Uganda; so-called 'iron beans' in Rwanda to reduce anaemia in pre-school children and women; and an African banana that is resistant to devastating wilt disease courtesy of a gene transferred from the green pepper.[54] I am often told that genetic engineering means monoculture by definition, and will therefore eliminate crop diversity – but there is no logical reason why this must happen either. The International Institute of Tropical Agriculture, for instance, which is behind the disease-resistant transgenic banana, also works to support the biodiversity of traditional crops: in September 2010 it helped launch a 'yam bank' in Nigeria to collect, store and preserve some 3,000 varieties of this important subsistence crop. The two strategies can go together.

No wonder plant biologists often appear so baffled by the controversy surrounding GE crops. 'It is hard to find good scientific reasons why this technology has not been universally embraced,' complain biologists Maurice Moloney and Jim Peacock in the academic journal *Current Opinion in Plant Biology*.[55] As they point out, 'All of the widely publicized objections ... have been soundly rebutted and relegated to the status of "urban myths".' The reality is that 'this technology has fundamentally changed agricultural production for the better. Yields are uniformly higher, and there is a dramatic decrease in the use of less-desirable pesticides,' amongst many other environmental benefits. As a major report by the US National Science Academy concluded in 2010: 'GE crops have had fewer adverse effects on the environment than non-GE crops produced conventionally.'[56] There may also be serious opportunity costs to the anti-GE furore preventing the uptake of nutritionally improved crop varieties: the transgenic 'golden rice', which has a higher Vitamin A content to help prevent blindness in developing-country children, is still stuck in the laboratory ten years after it was made ready for widespread deployment in Asia.[57] The barriers to deployment are not technical but regulatory – largely the product of many years of bitter campaigning by environmentalists. It is time for a change of tack by the Green movement, for the benefit of farmers, consumers and the environment.

Most importantly for the subject of this chapter, rejecting plant biotechnology means forgoing the opportunity to make what may be the single greatest agricultural breakthrough that might allow us to both feed the world and meet the nitrogen planetary boundary at the same time. This holy grail of genetic research is – just as the anonymous commenter under my 2008 anti-GE *Guardian* article proposed – to find a way to engineer our most important food plants to do what legumes manage to achieve naturally: fix their fertiliser direct from the atmosphere. This would enable crops to produce only as much nitrogen as they need, rather than farmers having to spread reactive nitrogen liberally and wastefully on the soil surface. But achieving this breakthrough, even with a technology as powerful as genetic engineering, will be far from easy. For a start, even legumes don't actually fix their own nitrogen in their tissues: the task is performed by microbes living symbiotically in a plant's specially created root nodules. To complicate things further, these nodules only work because they can somehow maintain the anaerobic (oxygen-free) conditions the microbes need to perform their nitrification miracle. At the moment, even the best plant scientists speak about getting nitrogen-fixation genes into major food crops as 'at least a ten-year effort'.[58] It may take even longer.

It seems almost impossible, right now, to imagine ways in which scientists could accomplish the trick. But then it seemed equally impossible, back in the days of William Crooke, to imagine that 'through the laboratory' the starvation that seemed to threaten at the end of the nineteenth century could be turned into a century of plenty. But back then the optimists were right and the pessimists wrong. Creating new strains of rice, wheat and corn that fix their own nitrogen could achieve in the twenty-first century what the Haber-Bosch breakthrough managed for the twentieth, and without the serious environmental drawbacks of industrial ammonia production. Environmentalists should not be scared of this prospect; they should welcome it. There can be no more important task than feeding people whilst protecting the planet. We must use the best of science and technology to help us to achieve this vital aim.

CHAPTER FIVE

The Land Use Boundary

Consider humanity's most serious planetary impacts covered so far. From climate change to biodiversity loss to nitrogen pollution, our unwitting changes to the Earth system have been extensive, mostly damaging and often irreversible. But they have also been largely subtle, long-term and imperceptible, and therefore easy to ignore or deny. The subject of this chapter, on the other hand, is not at all invisible – because it is taking place right in front of our eyes. No one can plausibly deny the utterly transformative impact we have had on the land.

The vast majority of the planet's ice-free land surface – 83 per cent according to one study[1] – is now influenced by humans in some way or another. This may seem surprising, shocking – or perhaps merely stating the obvious. Where I live, in the British Isles, no part of the landscape is totally unaltered by people. Even 'wild' areas like national parks look very different today to their natural pre-human state. The Lake District, for example, if left ungrazed by sheep, would revert to dense woodland on all but the highest peaks. Throughout the entire United Kingdom, the only species that have survived into the modern era are those that are able to coexist with human domination of the land: others, from beavers to wolves, have been extirpated entirely.

Human impacts on land may be much greater than is obvious at first sight. Roads, for example, appear to directly affect only a relatively small strip of land, but they also cut ecosystems in half, altering the survival prospects of species living on either side of them. With an estimated 1 million animals killed every day on America's road network, the effect of this constant removal of predators and prey is felt over much wider areas.[2] A seminal 2002 study of the ecological effects of a busy four-lane highway in Massachusetts found impacts – varying from wetland drainage to noise – across a broad 600-metre corridor. The consequent nationwide effects over the United States' entire 6.2-million-kilometre road network can only be guessed at.[3]

Whilst busy paved roads are a recent phenomenon, general human transformation of the land surface has been accelerating for millennia. The Roman empire deforested large areas around the Mediterranean, contributing to soil erosion and declining fertility. Unsustainable land-use change has been associated with the collapse of entire civilisations, from Easter Island in the Pacific to the Maya of Central America, as Jared Diamond documents in his book *Collapse*.[4] The European continent's landscape changed dramatically between AD 500, when it was still four-fifths covered by swamp and woodland, and AD 1300, when half of this natural land had already been converted to agriculture. This transformative process fluctuated in lockstep with human population growth: when the Black Death killed a third of Europe's population in the early fifteenth century, forests stopped their decline and began to regrow. To this day, many of Germany's most valued 'natural' woodlands owe their existence to the depopulation wrought by the medieval plague.[5]

Appearances can be deceptive: 'terra preta' (black soil) deposits along the Amazon river testify to large-scale cultivation and settlement by long-vanished civilisations, and suggest that large tracts of today's 'virgin' forest are probably regrowth across previously managed areas that were abandoned after diseases introduced by European invaders devastated the native population.[6] As with the English Lake District, national parks and other designated 'natural' areas around the world are mostly still under heavy human influence, particularly given the sheer numbers of people who visit them in search of recreation. The Fuji-Hakone-Izu Park in Japan is visited by 100 million people annually and includes spas, golf courses, hotels and trams amongst its supposedly natural landscape.[7] Even where minimised, our effects can sometimes be quite surprising: one US study found that birthing female moose use visitors to Yellowstone

National Park as 'human shields' by choosing calving grounds near roads, which traffic-averse predatory brown bears avoid.[8]

With so little of the Earth's land still pristine and unaffected by humans, the idea of the 'wilderness' has less and less meaning in the modern world. Indeed, if pollution and climate change are taken into account, no part of the planet's surface is any longer truly wild. This does not mean that we must gloomily accept the continuing diminution of semi-wild areas and the erosion of the vital ecosystem services they provide. It does mean though that we need to challenge some orthodoxies that are no longer useful in this new era of near-total human planetary dominance. 'Getting close to nature' or going 'back to the land' will generally not be good for the environment, however psychologically fulfilling these objectives may be to individuals seeking escape from industrial living. Instead, we need to intensify agriculture and other human land uses in existing areas as much as possible, and encourage as an environmental boon the growth of the world's major cities that already successfully concentrate today's enormous human population onto only a tiny proportion of the world's land. The most positive trend of all in allowing us to minimise our impact on the planet's surface is one more often bemoaned than celebrated: urbanisation.

LAND AND FREEDOM

Let us get one thing straight at the outset. The human transformation of land has been extremely good news for our species overall, allowing a population now approaching 7 billion people to be supported at ever-rising standards of living and comfort. Each of us now lives like a medieval king, offered choices of sumptuous foods from around the globe and surrounded by machines that wash our clothes and dishes at the touch of a button. I know many well-intentioned people who claim to despise the modern world for environmental reasons, despite themselves owning two cars, centrally heated houses and a dishwasher. I think we should celebrate what modern humans have achieved through industrialised civilisation, for its benefits in terms of quality of life are overwhelming for the vast majority alive today.

Often I find thinkers on the political right put this truth best. As the self-styled 'rational optimist' Matt Ridley – with whom I disagree about a great deal – writes: 'The vast majority of people are much better fed, much better sheltered, much better entertained, much better protected against disease and much more likely to live to old age than their ancestors have ever been.' He continues: 'This generation of humans has access to more calories, watts, lumen-hours, square feet, gigabytes, megahertz, light-years, nanometres ...' (the list goes on for a while) '... tennis rackets, guided missiles and anything else they could even imagine needing.' Nor is the global inequality frequently bemoaned by leftists much of a fly in the ointment: 'The average Botswanan earns more than the average Finn did in 1955. Infant mortality is lower today in Nepal than it was in Italy in 1951.'[9] And so on.

Where I disagree with Ridley and many of his colleagues on the right is their tendency to then go on to downplay or deny the environmental consequences of this human great leap forward, as later chapters will show. For Ridley in particular, his determination to show that the party can go on for ever has led him to stray into very unscientific stances on issues covered in this book like climate change and ocean acidification. Why not just admit candidly that whilst the human advance has been amazing and hugely beneficial, it has also had serious environmental impacts? As this book shows, Earth system boundaries can now increasingly be defined scientifically, and our job as modern, knowledgeable humans is to use all the tools at our disposal to avoid trespassing over these boundaries – at the same time as we must seek to allow the growth in human prosperity and numbers to continue for the foreseeable future.

For land in particular, our impacts on the planet's terrestrial surface are now so extensive and transformative that they threaten the capacity of varied ecosystems to self-regulate and maintain the living biosphere overall. Natural landscapes filter water, produce oxygen, sequester carbon, and provide vital habitat for other species. Whether these landscapes harbour forests, wetlands, grasslands or even deserts, the planet's varied biomes are an integral part of the Earth system – and are essential to its continued sustainable functioning.

Computer models of the Earth system can demonstrate how varied ecological zones are important to the planet overall. In one modelling study, climatologist Peter Snyder and his colleagues tried deleting entire biomes from their simulated planet to see what then happened to the climate. Remove the savannahs, they found, and you reduce rainfall totals by a third.[10] Press the delete key over the tropical forests, and entire continents see dramatic reductions in their precipitation and sudden rises in temperature. Different biomes, from temperate forests to the Arctic tundra, likely have significant effects on winds, temperatures and atmospheric circulation patterns.

As more and more of the Earth's surface sees these traditional biological zones transformed into 'anthropogenic biomes' dominated by humans, Earth-system scientists expect the planet's functioning to become steadily more unstable. Across at least three-quarters of the world today, humans are as or more important in determining which species make up the ecosystem than previously dominant factors like latitude, climate and geography.[11] Instead of the neat green and yellow strips that we learned about in geography class, denoting jungle, desert and so on, the planet is now divided into a complex mosaic of different human-influenced ecosystems. Human capture of the planet's productive capacity, according to the most recent relevant study, now adds up to about 24 per cent of the total productivity of the Earth's terrestrial biosphere.[12] In other words, a quarter of everything produced by all plants on land is eaten or otherwise consumed by us.

Different landscapes are exploited with different intensity: whilst forests will yield up to a fifth of their annual production in fuel, fibre or timber, cropland allows us to grab an impressive 83 per cent of the yearly productive share per hectare. The human species, therefore, whilst comprising only half of 1 per cent of the global animal biomass, consumes a significant fraction of everything the Earth produces.[13] This triumph for us is of course a disaster to the species we have displaced from their food webs: the disappearance of habitat and food supply is currently the greatest cause of the planet's continuing loss of biological diversity. It is also clear that ecological tipping points can be crossed if we push this process too far, with potentially irreversible consequences as overgrazed grassland tips into desert,[14] or as degraded tropical forest dries out and burns over vast areas of Indonesia and Brazil.

How far this trend can continue before precipitating some kind of regional or even global collapse in the functioning of the biosphere was considered by the planetary boundaries expert group, which concluded that no more than 15 per cent of the Earth's surface should be converted to cropland in order to protect the Earth system as a whole.[15] We are perilously close to this proposed planetary boundary already, with 12 per cent of land already devoted exclusively to agriculture. This leaves only 400 million hectares of additional land to be brought into production, a major challenge for humanity given population growth and increases in wealth and consumption across the planet. It is a challenge we can still meet, however – but only if we make the right decisions about how to use the planet's land most wisely.

A PLAN FOR LAND

It is not all bad news. Much of the world's land surface is already officially protected from damaging interference. At the last count, some 44,000 conserved locations covered 14 million square kilometres, encompassing nearly a tenth of the Earth's terrestrial area. These collectively represent, as one ecologist puts it, 'one of the most stunning conservation successes of the 20th century'.[16] Not all are properly managed, unfortunately – more than 70 per cent of 200 parks in tropical countries are affected by poaching, land invasion, logging and a variety of other threats,[17] and few have the resources to tackle them effectively. Priority number one for effective management of the planetary surface has to be to properly protect remaining natural or semi-natural areas, however degraded they may already have become.

This need has already been noted at the international level. At the 2010 meeting of the parties to the Convention on Biological Diversity in Nagoya, Japan, nearly 200 countries agreed on a target for at least 17 per cent of terrestrial and inland water areas to be managed in 'ecologically representative and well connected systems of protected areas and other effective area-based conservation measures' by 2020, by which time 'the rate of loss of all natural habitats, including forests' should be at least halved and 'where feasible brought close to zero'. Whilst not legally binding in the sense that sanctions will be taken against countries that fail to meet them, these targets at least represent a step-change in the attention given to biodiversity and land use by national governments. The Nagoya agreements represent a huge opportunity for conservationists: now that international targets have been set, we must force policymakers to take them seriously, and to translate aspirations into real, practical action that protects ecosystems in situ.

We also need to better highlight successes where they exist, rather than constantly lamenting failure. This is good psychology as well as good politics: people are naturally more motivated to act when they see that in other areas similar actions have yielded environmental benefits. Only a few of us gravitate towards lost causes, and then only until disillusion sets in. In Brazil, the government of former president Luiz Inácio Lula da Silva has successfully reduced deforestation in the Amazon by three quarters from a high in 2004, achieving something many conservationists had never dreamt they would see. In October 2010 President Lula announced that Brazil's target of an 80 per cent reduction in its deforestation rate in the Amazon could be met as early as 2016, four years ahead of schedule.[18]

This is good news for the whole world's climate, as billions of tonnes of carbon are preserved in forests and soils rather than being released into the atmosphere. But protecting the Amazon rainforest is also a win–win strategy locally, as large areas of healthy forest are necessary to maintain the wet climate that the rainforest needs to survive overall. Brazil shows that strong government policies can make a real difference: in 2000 less than a tenth of Amazonia was protected, whereas since then conservation areas on both federal and state lands have increased five-fold to more than 1.25 million square kilometres, covering nearly a quarter of the land area of the forest.

An important part of this strategy is that the government is working with indigenous people who live in the Amazon region: studies have shown that once granted land rights, local tribes have successfully protected huge areas against deforestation, yielding enormous benefits for biodiversity and the climate.[19] According to one computer-modelling study, state parks and indigenous lands – which together add up to 2.3 million square kilometres, or 43 per cent of Amazonia – may already be sufficient to buffer against a climatic tipping point that would dry out the forest.[20] This is no argument for complacency, however: in the Amazonian state of Mato Grosso, the state legislature recently passed a bill that would strip away much of the rainforest protection and increase the area zoned for cattle ranching and soybean plantations.[21]

At a global level, it matters a great deal which bits of land are ploughed up and which are protected. The 17 per cent of land identified as a target by the Nagoya agreements to be conserved must be zoned in line with the biodiversity 'hotspots' that harbour the vast majority of the world's threatened species. An incredible 44 per cent of plants and 35 per cent of animals are confined to 25 hotspots covering only 1.4 per cent of the land surface of the Earth.[22] This is an enormous opportunity: we can conserve nearly half the world's biodiversity by protecting just over 1 per cent of the total land area – a fantastic bargain by any assessment. One of the problems of today's parks and reserves is that they do not necessarily correlate with areas of high biodiversity or unique and threatened species: indeed, many areas of high conservation value are not protected at all – their defenceless inhabitants are consequently known as 'gap species'.[23]

The Nagoya targets represent a bare minimum, however. We must also seek to protect as much as possible of the world's last true great wildernesses, from the North American deserts to the mountainous forests of New Guinea, not just for their biodiversity value but also because these are the places that most closely approximate the natural state of the planet's biomes, and are thus likely to be essential components of the Earth system. The knowledge that true wilderness remains and will continue to do so also has great cultural and psychological importance for people, even city-dwellers who rarely if ever visit wild areas. How many of us ever see the Arctic tundra? Yet I am sure an overwhelming majority of us gain a benefit from knowing it is there. One recent study concluded that 44 per cent of the Earth's surface remains relatively wild, particularly the Northern Hemisphere boreal forests, Arctic tundra areas in Siberia and Canada, the great deserts in North Africa, Central Asia and Australia, and the tropical forest belt. These areas remain 70 per cent intact even today, and are sparsely populated with just 1.1 people or less per square kilometre (excluding urban settlements).[24]

However, just 7 per cent of this area is currently protected. A second urgent priority for the international community is to recognise the global value of wilderness, and to devise protection strategies that compensate the 80 million or so rural inhabitants living in these areas for the opportunities forgone by the need to curb disturbances like grazing, logging and mining. Given that these areas mostly have low if any agricultural value (no one is suggesting ploughing up Antarctica, for obvious reasons), the bill for global protection could be as low as $50 billion.[25] If this one-off payment is concentrated as an endowment, then management and compensation could be provided in perpetuity from the interest it generates. Here is an opportunity for conservationists to be much more ambitious, and to focus on vast areas of the planet that have little economic value and so can be protected very cost-effectively. We need a global movement for wilderness protection that is less focused on small areas within nation-states, and more concerned with transboundary ecological zones that extend around the entire planet.

Hand in hand with the protection of high-biodiversity and wilderness areas must come a better use of the land that is devoted to human-centred agriculture. This does not mean, as the previous chapter showed, that world farming should become organic: quite the opposite. Because organic agriculture produces on average only half the yield of crops per unit of land as conventional farming, any mass conversion to organic would end up using much more land and rapidly breaching the land-use limit proposed by the planetary boundaries expert group. Here some tradeoffs between different planetary boundaries come into evidence: organic is certainly better for reducing nitrogen runoff and the toxic impacts of pesticides, and may also help to promote wildlife in situ; but by using more land it is bound to end up further displacing natural ecosystems and reducing biodiversity overall. Instead, intensification – producing more yield per unit of existing farmland – using advanced farming technologies and high-yielding varieties of crops offers a more promising route to feeding a larger and more prosperous human population. Even so, some additional land will need to be brought into production to satisfy future increases in demand. The planetary boundaries expert group suggests a focus on abandoned cropland in Europe and North America, together with land in the former Soviet Union and 'some areas of Africa's savannas and South America's cerrado' for this unavoidable increase in cultivated area.

Questioning organic agriculture because of its land-use implications does not mean rejecting ecological principles in general. Instead, it means that farmers need to take the best of modern science and ecology to deliver maximum yields with minimum environmental damage. The precise combinations of herbicides, pesticides, irrigation, machinery and crop types will differ in different regions and climates. Some benefits may be unexpected: as we have already seen, the combination of weedkillers and genetically engineered herbicide-tolerant crops has allowed the widespread adoption of no-till agriculture in the US and Canada, improving soil carbon stocks and reducing erosion. Transgenic crops that incorporate their own pest resistance have also dramatically reduced the use of toxic pesticides, benefiting the environment still further.

In some areas, integrated pest management – encouraging natural predators to reduce pest infestations in crops – may be a better way to maximise yields than toxic pesticide sprays, and could work well together with genetic engineering. Combining different crops can also be highly beneficial, for example via agro-forestry, where trees are part of the production system and deliver shade and timber as well as benefits for wildlife. Aquaculture, where fish are raised in water-storage ponds or in flooded rice paddies, help recycle nutrients as well as maximising the production of animal protein on a farm. The agricultural scientist Jules Pretty calls this 'sustainable intensification' – maximising productivity whilst also protecting ecosystems as far as possible within the wider landscape.

Just down the road from me in Oxfordshire, outside the pretty village of Wytham with its thatched stone-built cottages, is the Food

Animal Initiative, a farm run on ecological principles that seeks to meet these challenges by combining the best of modern science and traditional knowledge. Walkers in the FAI's fields on the banks of the Thames will find sheep and chickens sharing space under specially planted shade trees, whilst barn owl nest boxes scattered around the farm encourage the population of this threatened farmland bird. Nearby haymeadows, protected as Sites of Special Scientific Interest, nurture bee orchids and other rare grassland flowers, whilst carp are raised in a former slurry lagoon that now collects water runoff from the main farm buildings. FAI has a successful outreach programme to help youngsters appreciate where their food comes from: my son Tom and his classmates from the local primary school enjoy their fortnightly visits to the farm and learn a great deal about ecology, the seasons and farming in the process. FAI's eggs are sold in our village shop as well as further afield. None of this means rejecting modern technology or big business: one of the farm's partners is McDonalds Europe, and FAI's director Mike Gooding makes clear that ecological farming must be 'commercially robust' if it is to be widely practised and effective.

MEAT AND ENERGY

There are two major current trends that mean the world must continue to produce more and more food from a limited area of land, probably necessitating a doubling of production within forty years. The first is the growing world population, which will reach 9 billion or more people by mid-century. The second is the tendency of more prosperous populations to consume more food in general, and to increase the proportion of meat and dairy in their diets. It is important to recognise that there is almost nothing we can or should do to influence either of these trends inasmuch as they affect the environment. Population growth is a given for now, as I will discuss later. Moreover, efforts by well-intentioned Westerners to convince richer Chinese people not to eat more meat are likely to meet with a robust response, to say the least. There is certainly a strong health case to be made that people in developed countries, the US in particular, currently consume too much meat and saturated fats. But campaigners are on to a loser if they try to convince people either to convert en masse to vegetarianism or to have fewer children for the good of the planet.

People's desire to eat more meat as they get more wealthy is so deeply embedded in most cultures (and getting lots of protein may even be a biological impulse inherent in all of us) that it is not something that is amenable to outside influence. As with climate change, the only pragmatic option is to concentrate efforts to fulfil people's desires and demands in a way that protects natural ecosystems as far as possible – not to try to challenge patterns of consumption per se by insisting that they are unsustainable, even if this appears to be the case in the short term. Such an approach has failed in the past and will continue to fail in the future. It may also be counterproductive to insist that people deny themselves consumptive pleasures or offspring, sparking populist campaigns that seek to deny the very reality of environmental challenges. A successful environmental movement must work with people's aspirations for prosperity and comfort, not try to suppress these impulses.

One trend that can and should be challenged however is the global rush to biofuels. There is nothing inevitable about the increasing use of corn to produce ethanol or soybeans to manufacture biodiesel – these technologies offer no benefits to consumers, and have instead been driven by mistaken subsidies supposedly aimed at tackling climate change, but which instead most likely make it, and a whole host of other environmental problems, worse. I once met a well-known American environmentalist who proudly told me that he burned wheat on his home biofuel stove to keep his house warm. I was horrified – and I still am. Putting ethanol from US corn into cars is equally wasteful, unethical and unnecessary. Burning food crops for power is the worst use of scarce land imaginable, and has already led to a situation where there is a direct conflict between food and energy: a significant proportion of the food-price spike in 2008 (and a further spike in early 2011), which led to widespread hunger and bread riots in many poorer countries, was driven by crops being withdrawn from international markets to produce biofuels for transport.

Currently more than 40 per cent of the US corn crop goes into producing ethanol, which is mostly mixed with gasoline to fuel conventional cars.[26] Even though this raises food prices and reduces the US surplus that can be sold on world markets, the ethanol industry is supported by a $6 billion subsidy scheme aimed at cutting greenhouse gas emissions. In Europe, biofuels are similarly supported with 'environmental' subsidies and fuel mandates. All these supports and promotion schemes for liquid biofuels should be scrapped, because of overwhelming scientific evidence that using land to produce energy crops delivers no climate benefits at all once agricultural emissions and land-use change are taken into account. This is obvious in the most egregious example of all: the clearing of tropical rainforest in Malaysia and Indonesia for oil-palm plantations, at least a third of which are used to produce feedstocks for biofuels (the rest goes into processed food, from chocolate to cooking oil, and cosmetics). Every year vast areas of highly biodiverse rainforest are being felled or burned in the Indonesian islands of Borneo and Sumatra for conversion to oil palm; the devastation has been highlighted by Greenpeace and many others.[27] Overall, 1.7 million hectares of Indonesian forest were converted to oil-palm plantation between 1990 and 2005,[28] and the rate of destruction is accelerating. Scientists have calculated that the burning of peatland rainforest to free up land for plantations churns out more than 1,500 tonnes of carbon per hectare.[29] Mopping this CO_2 up using palm-oil-derived biodiesel would take nearly 700 years. Biofuels derived from cleared rainforest land should not just be discouraged – they should be outright banned.

Estimates of future land take for biofuels production range up to well over a billion hectares globally,[30] more than double the 400 million hectares that remain if we are to respect the proposed planetary boundary. Once the need to produce more food is taken into consideration, it is clear that biofuels can only ever be a marginal contributor to world energy supplies. This conclusion has important implications. Most critically, liquid fossil fuels used in the world's vehicle fleet of cars and trucks cannot simply be replaced with liquid biofuels. Instead, surface transport must be almost entirely converted to electricity. This is starting to happen already: all-electric cars or plug-in hybrids are beginning to hit the mass market, and will be particularly appropriate for urban or suburban drivers where the limited range of current battery technology is less of a concern. In the medium to long term, however, countrywide electrical-charging infrastructure will need to be built that allows electric cars to deliver all the range and refuelling convenience of petrol vehicles. A US-based company called Better Place has already signed contracts in Israel, Denmark and other countries to pioneer battery-swapping stations that will allow motorists to swap dead batteries for fully charged ones in less time than it once took them to fill their tanks. If batteries can be designed that charge faster and last longer, most of us could simply plug in our cars at home or whilst parked in town – a much better option than driving to a petrol station and having to queue to pay afterwards.

Electric is clearly the way to go for the majority of surface transport. This includes mopeds and bikes as well as large trucks. With oil prices rising and local air pollution worsening in developing-world megacities, the tipping point may come even sooner than many pundits think, and be driven by demand in fast-growing countries like China. All the major automotive companies are now positioning themselves to exploit this massive future market: Nissan, General Motors, Toyota, Volkswagen, Honda, Ford, BMW, Tesla Motors and Daimler are already or soon will be offering affordable electric vehicles. Tesla's Roadster is a far cry from the electric vehicle's milk-float caricature: this all-electric sports car can race from 0 to 60 miles per hour in under four seconds, and has a top speed of 125 mph. And it is not just land-based transport that will probably go electric: electricity might also soon be the fuel of choice for small boats, which can more easily handle weighty stacks of batteries than cars, and where the strong torque provided by an electric motor could improve speed and manoeuvrability.

The only likely exception to the rule against biofuels is the urgent need to decarbonise air transport, where low-carbon alternatives to liquid hydrocarbon fuels remain a distant prospect. Whilst aviation has been demonised by environmentalists (myself included) in the past because of the climate-change impact of aircraft emissions, in terms of fuel efficiency per passenger kilometre the latest large aircraft like the Airbus A320 and the Boeing 787 now compare favourably with small family cars. The reason why per capita emissions from an intercontinental flight are counted in the many tonnes of CO_2 is the enormous distances covered: no one drives from London to Sydney. Reducing aggregate demand is not an option: pleas by Greens for people to 'give up flying' have found limited appeal to say the least, particularly given that most environmentalists I know continue themselves to enjoy the benefits of air transport.[31]

Therefore, with over 2 billion people using air travel every year already, and rapid uptake in developing countries like India and China, technical substitutes for high-carbon aviation must rapidly be found. If they can be sourced fairly sustainably, biofuels look promising, particularly 'second-generation' biofuels like algae that do not directly compete with food crops.[32] British Airways has led the way with its pioneering commitment in February 2010 to build a plant in the UK that will convert 500,000 tonnes of waste material into 16 million gallons of jet fuel annually.[33] This may seem like a large amount, yet it represents only about 2 per cent of flights from London's Heathrow Airport. This is the scale challenge of aviation, and demonstrates why biofuels may need to be almost exclusively reserved for air transport – surface transport must go electric.

However, where the electricity comes from to power the next generation of vehicles is of course vitally important. Electric cars charging up using coal power will deliver little or no benefit to the climate. Power stations using biofuels on a large scale, and gobbling up millions of hectares of land in the process, will be similarly disastrous. The best solution, as I showed in the climate chapter, will be a dramatically upscaled combination of renewables and nuclear, with the proportion of different technologies varying place by place. For the purposes of this chapter, however, it is the land-use implications of different energy options that are at issue – and here it is indisputable that nuclear wins hands down. The reason why is basic: renewables work by harvesting diffuse energy like sunlight and wind over large areas, whilst nuclear fission delivers prodigious amounts of energy from tiny amounts of source material. Compared weight-by-weight, uranium 235 delivers a million times more energy than coal, which itself already represents chemical energy in a highly concentrated form. Just how much energy nuclear fission releases is described by Einstein's famous equation $E=MC^2$, where E is energy, M is mass and C the speed of light, about 300 million metres per second. Clearly even with a very small amount of fissionable material, multiplying it by the square of 300 million yields a very big number.

Anti-nuclear campaigners try to magic away this famous piece of physics by arguing that uranium mining and the disposal of radioactive waste use up so much land as to make the comparison meaningless. They are mistaken. Even on the basis of a full life-cycle analysis, nuclear uses much less land than solar photovoltaics (PV) and wind.[34] (Biomass comes out worst of all, using more than a thousand times the land area of nuclear power.) The energy efficiency advocate and dogged anti-nuclear activist Amory Lovins argues that wind uses land *more* efficiently than nuclear, but only by entirely ignoring the land in between turbines, which have to be sited a substantial distance apart in order not to interfere with each other.[35] Whilst it is true that some cultivation or grazing can go on underneath wind parks, to consider the land outside the actual concrete base of the windmills as equivalent to an undisturbed ecosystem is perverse. One could equally argue – and many anti-wind campaigners do – that the land-use impact of wind farms is much wider than their physical footprint because of their visual intrusion on the wider landscape, and the wildlife impact of birds and bats killed by spinning turbine blades.

The latter issue suggests a possible conflict between the climate change and biodiversity boundaries as renewable energy production is scaled up, as it surely must be. Although I hear many clean-energy enthusiasts downplaying the wildlife impacts in their understandable enthusiasm for green power, there is no doubt that wind farms in some areas do kill significant numbers of birds. One recent study of bird mortality in California's 5,400-turbine Altamont Pass Wind Resource Area found an annual death toll that included 67 golden eagles, 188 red-tailed hawks, 347 American kestrels, 30 barn owls, 440 burrowing owls, 34 mallards, 120 mourning doves, 189 rock doves, 271 starlings, 415 western meadowlarks and many other birds too numerous to list here.[36] With this kind of mortality rate every year, large wind parks like this one could be significant population sinks for birds.

Whether this matters partly depends on the species. Bird carcasses found around wind turbines on Ontario's Wolfe Island in the first half of 2010 included, according to local reports, an osprey and seven red-tailed hawks.[37] The latter death toll may comprise more than a third of the local population, whilst the single osprey could be a tenth of the population of that species, suggesting that this particular wind farm could be affecting rare raptors in particular. However, the number of birds killed varies dramatically between wind parks, depending on topography, surrounding habitat and local wildlife population. There is also some evidence that newer, larger turbine designs are less deadly to birds. However, these taller windmills are now thought to be a significant threat to bats, because they extend up into the airspace used by bats whilst they migrate. Again, location matters a lot: in a survey of wind-farm mortality studies across North America, some turbines killed no bats at all, whilst others killed 20–30 each.[38] An April 2011 assessment published in the journal *Science* warned of more severe impacts, however. According to zoologist Justin Boyles and his team, wind energy parks could join the deadly 'white nose syndrome' fungal pathogen in wiping out entire bat populations across the United States, causing $3.7 billion in agricultural losses due to bats no longer consuming insect pests.[39] An earlier assessment of future bat mortality projected that up to 111,000 bats could be being slaughtered annually by 2020 as wind farms sprout across the forested ridgetops of the eastern United States.[40]

On the other hand, properly sited wind parks may be quite benign in terms of their wildlife impacts, and some are being designed to actually enhance the biodiversity value of their impacted area. In South Wales, for example, Nuon Renewables is proposing an 84-turbine wind farm in a large area of upland peat bog that has been significantly damaged in the past by inappropriate government-planted coniferous woodland. Nuon is proposing to combine with its wind farm the largest peat-restoration project in South Wales, and is focusing its efforts on protecting and encouraging the local population of the rare honey buzzard by siting turbines away from its hunting areas and increasing the areas of suitable woodland habitat nearby.[41] But the issue of scale will be very important for onshore wind in general: optimistic projections from climate campaigners suggest upscaling wind parks by a hundred times or more in the US, UK and elsewhere. Unless some way is found to significantly reduce bird and bat kills, this level of wind generation could severely conflict with the biodiversity boundary.

It is not only wind that is at issue. In the UK the environmental movement is deeply divided over long-standing proposals to dam the Severn Estuary with a tidal barrage. Because it could potentially supply 5 per cent of Britain's electricity in a zero-carbon way, the idea was supported by the Sustainable Development Commission in 2007, headed by former Friends of the Earth director Jonathon Porritt. But the barrage proposal was vociferously opposed by Friends of the Earth, the National Trust and the RSPB (Greenpeace remained on the fence) on the grounds of its wildlife impacts. Here, I would support Friends of the Earth and its allies. Habitat loss is the greatest threat to UK biodiversity. According to research published in 2010 by Oxford University ecologist Clive Hambler and colleagues, despite all its wildlife-protection laws 1,000 endangered species still stand on the brink of extinction in the UK, with an estimate of one species going extinct in England every fortnight.[42] Mudflats and wetlands such as those that would be damaged or destroyed by the Severn Barrage are vital habitat for endangered waterfowl, as well as fish and many other rare species, and are among the most threatened habitats in the country.[43] 'To use climate change as an excuse to ignore so many ecologists would be eco-lunacy, especially since renewables are not the only option,' complained Hambler in a strongly worded letter to the *Independent*.[44] Unlike Friends of the Earth and Jonathon Porritt, however, Hambler acknowledges that nuclear is a better option in this case. 'Despite popular belief, nuclear power is highly acceptable from a conservation perspective,' he writes in a recent book.[45]

My argument is not in favour of nuclear and against renewables, however – both are necessary, in soaring quantities, if we are to meet the climate change planetary boundary. As Britain's energy secretary Chris Huhne rightly put it when rejecting public funding for the Severn Barrage: 'I'm fed up with the stand-off between advocates of renewables and of nuclear that means we have neither.'[46] Both zero-carbon energy options, and various others, are essential to tackling climate change. Moreover, renewables can often be deployed in ways that minimise land-use and biodiversity impacts. Where building-mounted solar photovoltaics are cost-effective, no additional land – and therefore wildlife habitat – is used at all. Even the energy-hungry US could probably supply all of its electricity use, according to one study, using solar PV on an area equivalent to 0.6 per cent of the country – just a fraction of the existing developed and urban space.[47]

The objections to this are not environmental but economic: solar PV is still extremely expensive, and only commercially competitive in places where fossil-fuel alternatives are costly or where subsidy schemes like feed-in tariffs support the nascent solar industry. Whilst aggressive solar subsidy schemes in cloudy northern countries like Germany probably do not deliver value for money, in countries with high levels of sunlight solar power must eventually become the largest generator of electricity. As I argued in the climate chapter, North Africa should and probably will become a major supplier of desert solar-generated power to Europe, whilst Australia could easily become the world's first major 100 per cent solar country if the government in Canberra were prepared to show greater leadership than it has so far on climate-change mitigation. Feed-in-tariffs, which pay generators a premium for solar-generated electricity, may be the best policy option to drive this transition, and have helped dramatically increase the use of solar PV in countries like Spain, Germany and Italy.

Solar thermal plants – which concentrate the sun's heat directly to drive a central steam turbine generator – will use a lot of land because they need thousands of mirrors to be sited in the same place. This may not matter much in the northern Sahara, where the biodiversity impacts of massive-scale solar deployment are likely to be minimal in the hyper-arid desert. But in the US southwest, conflicts are already appearing between solar enthusiasts and conservationists. On 30 December 2010 the Sierra Club filed a suit with the California

Supreme Court to stop a massive solar plant going ahead in pristine habitat in the Mojave Desert, claiming that threatened species like the desert tortoise, Mojave fringe-toed lizard and even the golden eagle would be harmed.[48] Earlier the same month, a coalition of Native American and civic groups filed a legal suit against the US Department of the Interior regarding six different solar thermal projects, aiming to save not only desert habitat but also cultural landmarks important to the tribes. Here I would side with the conservationists. The Mojave Desert is one of the great wildernesses mentioned earlier in this chapter, and is particularly fragile given its location near massive population centres in the highly developed US southwest. It should be properly protected as important wildlife habitat against all forms of development – renewable energy included.

The issue is a classic illustration of the challenges that climate-change mitigation will throw up regarding land use – the biggest of which is the danger of what a group of US conservation experts term 'energy sprawl'. The conservationists, working with the American group The Nature Conservancy, have quantified the energy-sprawl implications of different low-carbon technologies, and their results make challenging reading.[49] Supporting what I argued earlier, biofuels come out worst by an enormous margin, gobbling up 300–600 square kilometres per terawatt-hour of electricity produced each year. At the other end of the scale, nuclear is least land-intensive with only around 2 square kilometres affected to produce the same amount of power. Renewables are in the middle, with solar thermal needing 15 square kilometres, solar PV (on open land rather than on buildings) 36 km^2, and wind 72 km^2.

Luckily there are many ways that the land-use intensity of renewables can be reduced. As I have already pointed out, solar PV can be mounted on existing buildings and other developments. For wind, the obvious solution to the land-use conflicts of 'energy sprawl' is to base most of the industry offshore. The further offshore the better: whilst studies have found habitat-reduction impacts on ducks and other seabirds for shallow-water turbines, there are many fewer species further away from the coast.[50] Technologies are already being developed for deep-sea floating turbines, with each windmill potentially delivering as much as 10 megawatts of power – enough to provide electricity to several thousand average Western households.

In the shorter term, countries with large areas of shallow continental shelf are building large offshore wind parks with turbines anchored to the seabed. For the UK much of the North Sea is available for wind development, and the country is already leading the world in offshore wind deployment – the theoretical ultimate resource potential is many times Britain's total current electricity consumption. With its windy coasts, the Scottish government in September 2010 announced an impressive target of becoming 80 per cent renewable within the next decade.[51] In Europe things are moving fast: in December 2010 ten European countries signed an agreement to develop an offshore electricity grid in the North Sea, the first stage in a continent-wide 'supergrid' that will be necessary to balance out the intermittency of renewables between different countries and deliver clean power to faraway population centres.[52]

Of course, land use is not the only consideration when assessing the merits of different energy options. Coal and gas come out quite well in terms of their land intensity in the 'energy-sprawl' study (coal uses about 10 km^2 of land per terawatt-hour of electricity produced each year, and gas about 18), but both must be phased out because of climate change unless their carbon emissions can be captured and buried underground. Hydroelectric reservoirs also take up little space, but have impacts on river ecosystems far downstream, as the following chapter will show. At the risk of repeating myself, in terms of the land use planetary boundary alone, my conclusion is that nuclear power is likely to be the most environmentally friendly technology of all, although appropriately sited wind, solar and other renewables are similarly benign and should be equally encouraged.

REDD OR DEAD?

Traditionally land-use planning is the poor relation of other environmental considerations, left to the mercies of local councils or town boroughs. But with a planetary boundary proposed to limit the human use of the globe's terrestrial area, we need to start thinking about zoning at a global scale. In some countries there is little wilderness or area of high biodiversity left to protect. In others, however, habitats of worldwide importance remain almost unscathed. Many of these countries are still developing, and for them to forgo the development potential of ploughing up grasslands or levelling old-growth forests – which yield undeniable short-term economic benefits in any conventional analysis – is a key opportunity cost that will be borne disproportionately by poor countries. If the rainforests are to be left standing for the benefit of the entire world, therefore, the entire world should pay for them – or at the very least the burden of the opportunity cost should be more fairly shared.

Ecuador illustrated the potential conflict of interest by offering in September 2010 to forgo the opportunity of developing new oil wells in its Amazonian forest. In return for not drilling in its Yasuni National Park, the country's president has asked for international contributions into a compensation fund set at $3.6 billion.[53] Elsewhere, Norway's government has been a leader in international tropical forest protection: in May 2010 it offered $1 billion to the Indonesian government in return for a moratorium on new forestry concessions to logging and palm-oil companies.[54] Norway has pledged another $1 billion to the Amazon Fund, specially created to reduce deforestation in Brazil, whilst both the UK and Norway have jointly committed $200 million to a Congo Basin Forest Fund.[55]

Whilst bilateral deals such as these are an important first step, clearly a global system is needed to pay for forest protection. Just such a scheme, termed Reducing Emissions from Deforestation and Degradation (REDD), is under discussion as part of the international climate negotiations. The idea is that by bringing forests into world carbon markets, enormous funds can be raised through international offsets to fund forest protection as a way of reducing carbon emissions. Getting REDD agreed and implemented is critically important, not only because tropical forest destruction is a major source of carbon emissions (amounting to 1.2 billion tonnes a year at the last count[56]) but also because of the need to preserve their biodiversity and importance at an Earth-system scale as a biome of planetary importance. Reducing emissions from deforestation therefore actually matters far more than shutting down coal-fired power stations, because tropical forests are treasure troves of biodiversity, whilst power stations are just concrete and steel.

Unfortunately many environmental lobby groups have been rather sceptical of REDD, and some have outright opposed it. 'During the climate talks, we will be demanding that forests are kept out of carbon markets,' vowed a Friends of the Earth (FoE) spokesperson in 2008.[57] Looking at the group's December 2008 pamphlet 'REDD myths',[58] one cannot help but get the impression that Friends of the Earth's opposition has more to do with a general aversion to trading and international markets than any specific objections to the REDD scheme. 'Allowing countries with carbon intensive lifestyles to continue consuming inequitably and unsustainably, by permitting them to fund cheaper forest carbon "offsets" in developing countries, diverts critical resources and attention away from measures to address fossil fuel consumption and the real underlying causes of deforestation,' FoE asserted. But by insisting that carbon reductions must for ideological reasons exclusively take place in rich countries, groups like Friends of the Earth risk leaving forests unprotected and their destruction to continue by default as they focus on wider political battles. Clearly if forests are to be worth more alive than dead, somebody, somewhere, will have to stump up the cash.

This in turn means that the incentive of potential profits must be central to any financing scheme – a prospect Friends of the Earth views with barely concealed horror. In a 2010 publication on the same issue, the group asserts in an outraged tone that 'many REDD projects are being set up specifically with a view to making a handsome profit', and that 'a REDD race is firmly underway, with investors, including banks, energy companies and carbon traders fully engaged in seeking out profitable opportunities'.[59] To my mind, that powerful interests are beginning to see forest protection as a winning opportunity should be welcomed by anyone who wants to see the survival of tropical forests in the long term. This is simply practical politics: no amount of glossy pamphlets published by Green groups will have a chance of protecting forests if moneyed interests stand to make a bigger profit from their destruction than their protection, as the accelerated rate of deforestation over recent decades has shown. Bringing forests into properly designed markets as living entities is essential for their survival.

Happily, the wider environmental movement has recently adopted a more realistic approach. At the December 2010 climate-change talks in Cancun, the Climate Action Network – a broad international coalition of Green groups – supported REDD during the negotiations, making constructive suggestions for how the draft texts could be improved. Perhaps partly due to this assistance, REDD principles were broadly agreed by the end of the conference, including safeguards for the rights of indigenous peoples, and the need for forest countries to begin setting up ways of monitoring forest loss and counting any reduction of emissions due to protection schemes. However, there remains the danger that until international carbon markets are strengthened there may be few places for forest-credit cash to be raised – and this in turn hinges on broader agreement about the future of the Kyoto Protocol and a wider carbon-reduction and trading effort bringing in the US, China and all the other countries that do not have Kyoto targets.

And REDD applies only to forests. At an even broader level, a global system is needed to pay for conservation in wetlands, grasslands, tundra, deserts and the other ecosystems of global importance. My proposal for how this might work is a straightforward one, based on the fact that the human economy depends completely on the 'natural capital' of a healthy biosphere. I suggest each country adds half a per cent to Value Added Tax (VAT) with the proceeds raised specifically safeguarded for ecosystem and habitat restoration ('rewilding') and preservation. This would be fair because a tax on consumption would mean that people pay in proportion to the environmental impact of their lifestyle patterns – with those who consume more bearing more of the cost – and as the economy grows so the yield from the tax would grow too. The amounts raised would be substantial, but barely noticeable to consumers: in the UK the rate of VAT has varied between 15 and 20 per cent in just a couple of years in response to the changing economic situation whilst raising little protest. In Britain, half a per cent extra on VAT would raise something in the order of £2.6 billion annually. My suggestion would be that a majority of this, and similar amounts raised in other developed countries, be put into an international fund to buy, endow or otherwise protect large areas of natural landscape and globally valuable ecosystems in developing nations.

SEX AND THE CITY

As a general rule – and making an exception for indigenous people and other communities who have demonstrated a long-term commitment to the sustainable use of their local environments – the fewer people who live in or close to rainforests and other important ecological biomes the better. Rural depopulation and urbanisation in developing countries are often decried by those who are concerned about the relentless expansion of megacities, which seem terribly unsustainable because of their noise, sprawling slums, congestion and pollution. But from the perspective of sustainable land use and habitat protection, the more that growing numbers of people can be persuaded to herd themselves into relatively small areas of urban land, the better for the environment.

Village life, particularly in extremely poor developing countries, should not be romanticised by outsiders. Whenever they are given the chance, younger generations tend to flee to the cities, where they have many more livelihood options and can escape the cultural oppression that is often a hallmark of traditional societies. In many parts of the world, if you want to marry the person you choose, be gay, be female and successful or avoid daily backbreaking labour carrying water or fetching firewood, then you probably need to move to the city. In 1975 there were just three megacities of over 10 million people. Today there are 21. It sounds scary, but this unstoppable shift towards urbanisation actually ranks as one of the most environmentally beneficial trends of the last few decades. As the UN Population Fund wrote in a recent report: 'Density is potentially useful. With world population at 6.7 billion people in 2007 and growing at over 75 million a year, demographic concentration gives sustainability a better chance. The protection of rural ecosystems ultimately requires that population be concentrated in non-primary sector activities and densely populated areas.'[60]

By the end of 2011 the world's population will stand at 7 billion. Seven billion people is an incredible number, but standing shoulder to shoulder we would all comfortably fit within the city of Los Angeles.[61] City living is seldom lauded by environmentalists, but it may be our most environmentally friendly trait as a species, because urban dwelling is vastly more efficient than living in the countryside. Shops and other services are more concentrated in town and city neighbourhoods, and urban residents are much more likely to use public transport, share heating and housing, and have lower carbon footprints than their rural brethren. But given the scale of global population growth, the challenge still seems daunting: the world will need to accommodate 2 billion more urban dwellers by 2030, a rate of expansion equivalent to building about 13 great cities (each with over 5 million inhabitants) per year, almost all in developing countries.[62]

But consider against this two pieces of good news. First the rate of population growth is declining, having fallen from 2.2 per cent at its highest in the early 1960s to 1.2 per cent today.[63] UN projections for peak absolute population numbers keep being revised downwards, and its 'low variant' projection has now come down to just 8 billion people by 2050.[64] In some countries the population is already falling in absolute terms: in Russia the overall population has fallen from 148 million to 142 million since the 1990s. Second, the amount of land space taken up by cities is actually relatively small compared with the number of people they shelter: satellite image composites show that urban sites cover only 2.8 per cent of the Earth's land; accordingly the UN estimates that about 3.3 billion people occupy an area less than half the size of Australia.[65] Imagine all these people forced to fan out into the countryside in some kind of Khmer Rouge-style Year Zero experiment: the result would be a global ecological disaster.

This gainsays conventional environmental wisdom in several ways. Clearly, the best strategy to curb future population growth is to speed up the 'demographic transition' in developing countries – and this transition towards women having fewer babies is inextricably linked both with increasing levels of prosperity and with urbanisation. Therefore rising rates of economic growth and the expansion of cities are good news for the environment because – more than anything else – they will restrain the future growth in human population. Moreover, although the idea of getting close to the land in small-scale communities has a deep cultural resonance in some schools of environmentalist thought, in reality this is probably the worst thing that anyone can do. It is much better to encourage as many people as possible to continue to seek to improve their livelihoods and prospects by embracing urban living and migrating out of rural areas.

All around the world, rural depopulation is leading to forest regrowth in abandoned areas – from the vast tracts of secondary broadleaf woodland in America's New England states to tropical forests in Puerto Rico,[66] the Dominican Republic and many other areas. In Costa Rica, abandoned cattle pasture is nurturing a flourishing young forest that in turn now supports a stable population of jaguars and other threatened fauna.[67] A recent scientific paper looking at Latin America lists 'similar patterns of ecosystem recovery following rural-urban migration' in Patagonia, northwest Argentina, Ecuador, Mexico, Honduras and the montane deserts and Andean tundra ecosystems of Bolivia, Argentina and Peru.[68] Even in rich countries, proposals for 'rewilding' – which I strongly support – only stand a chance of success in areas where rural populations have collapsed and formerly subsidised unproductive farms can be shut down to allow them to revert to nature.

These observations, and many other studies around the world, suggest that environmentalists need to take land use more seriously. This challenges much conventional wisdom, however. It suggests that rural depopulation should not necessarily be opposed with 'sustainable development' schemes aimed at improving rural life to stop people migrating to cities. Equally, instead of encouraging low-tech traditional farming methods it may be preferable to focus on improving high-yield mechanised agriculture on the most fertile farmland to feed the new urban residents, whilst allowing mountainsides and other marginal lands to revert to forest. This is already happening by default in Latin America and elsewhere: in Vietnam, forest area has been increasing since the 1990s after small-scale, unproductive agriculture was made uncompetitive by more intensive, larger-scale farming in the more open market economy. The environment has benefited as extensive areas were abandoned by people moving to take up jobs in the expanding cities.[69]

This trend should be cause for optimism that we can make progress in meeting the biodiversity planetary boundary. 'Current human demographic trends, including slowing population growth and intense urbanization, give reason to hope that deforestation will slow, natural forest regeneration through secondary succession will accelerate, and the widely anticipated mass extinction of tropical forest species will be avoided,' write the biologists Joseph Wright and Helene Muller-Landau.[70] Through examining UN data on forest cover and populations, Wright and Muller-Landau found a strong relationship between rural population density and deforestation, that appears to hold true across 45 developing countries in Africa, Latin America and tropical Asia.[71] The implication is simple: fewer people, more forest.

As always, one should not oversimplify. Cities themselves consume resources, including food, timber, water and energy, harvested over vastly wider areas than the land that they physically occupy, and this greater footprint needs to be considered in any overall assessment to get a true picture. When peasants move to the cities, their land might just as easily be turned over to large-scale cattle ranching or plantations as allowed to revert to forest.[72] Studies have suggested that this is particularly the case in Amazonia, where most deforestation is carried out by ranchers, so population density is less clearly linked with the fate of the forest.[73] Moreover, even after people move out, the recovery of forests cannot always be left to chance – it needs active management and ecologically friendly government policies.[74] Whether secondary forest can help avoid large-scale species extinctions also depends on the extent to which animals and plants accustomed to old-growth forests can successfully recolonise new areas.

But the overall conclusion seems to me irrefutable. Urbanisation is good for sustainability, because it reduces population growth and concentrates the overall human impact on the land in a smaller area. Handled properly, migration away from rural areas and into cities offers a huge opportunity for ecosystem protection and restoration. Our best hope for meeting the land use planetary boundary is therefore to encourage the trends towards rising prosperity and demographic transition in developing countries, in order to allow their forests and other important natural habitats to survive and regrow. Given the choice, most people around the world already find city life more attractive and varied than that in the countryside. Forget the 'back to the land' self-indulgence of some disgruntled people in rich countries. Billions of people want to move to urban areas to achieve increasing prosperity and improve their standard of living. Let us be glad of that. They are unwitting 'Greens', whose efforts at self-improvement should be celebrated.

CHAPTER SIX

The Freshwater Boundary

We have polluted the seas and appropriated land from other competing species – but water we have not so much stolen as imprisoned, behind concrete dam walls, within dark reservoirs and behind the high levees that hem in once-mighty free-flowing rivers like the Mississippi and the Yangtze. Whole natural drainage basins, which once responded to the grand seasonal cycles of winter flood and summer drought, now react meekly to the whims of water managers seated in the control rooms that govern sluice gates in tens of thousands of large dams. The Colorado River may have gouged out the most spectacular cutting in the world – the Grand Canyon – but today the flow of this powerful torrent is as much a product of human hydrological engineering as it is of any natural force. If you live in Las Vegas, Los Angeles or Phoenix, Arizona, the Colorado is more part of an enormous plumbing works that ends in your shower or bath, as it does for 30 million other people. Its ecological role has declined in tandem, for in an average year hardly a drop of the precious water in this 1400-mile river now succeeds in reaching the sea.

To put this in context: worldwide 60 per cent of the 227 largest rivers have been fragmented by man-made infrastructure, and the total number of dams blocking the natural flow of the planet's watercourses is estimated at 800,000.[1] These impound approximately 10,000 cubic kilometres of water – a quantity so substantial that it measurably reduces the rate of sea level rise (by about half a millimetre a year for the last half-century[2]) and even changes the mass distribution of the planet sufficiently to alter its axis and slightly increase the speed of its rotation.[3] The sheer scale of human engineering activity on rivers has been extraordinary: on average we have constructed two large dams per day over the last fifty years, half of those in China alone.[4] Humans have affected the water cycle in less visible ways too: deforestation and irrigation are altering water-vapour flows over the planet's surface;[5] changes in land use and climate are increasing total planetary river runoff.[6] Human engineering can have large-scale impacts – agriculture in Pakistan's dry Indus Basin, supported by the largest irrigation network of canals and dams in the world, probably has a direct effect on the region's monsoon.[7]

Due to the globe-girdling reach of modern human civilisation, these regional and planetary-scale changes are perhaps unsurprising, for humanity has always had an umbilical connection with rivers and freshwater. Imperial capitals throughout history have lined major watercourses, from Nanjing on the Yangtze to London on the Thames. When water became scarce or was misused – as in ancient Mesopotamia or during Central America's Classic Maya period – great civilisations could come crashing down, leaving little trace behind as their once-unconquerable cities were reclaimed by sand or forest. Today we face the danger of overusing water resources on a planetary scale, and the consequences for our advanced civilisation may be just as significant in the long run.

TURNING ON THE TAP

It would be foolish to neglect the enormous benefits that water engineering and control have delivered to humanity. Our domination of the water cycle, as with our rule of the land, has given the human species in the industrial era advantages that would once have been unimaginable: unlimited clean, fresh water, literally 'on tap' and delivered wherever and whenever we need it. For most of human history, and still today in poorer parts of the world, water had to be fetched and carried – usually by women – and was a precious and limited resource. Clean water could never be guaranteed, and as human populations began to rise after the Middle Ages, epidemics of water-borne diseases like cholera became a chronic danger for anyone living in the booming, crowded cities. That whole civilisations can now flourish in desert areas that experience little or no direct rainfall is a tribute to human ingenuity and the mastery of complex hydrological engineering over the vagaries of nature. That we have in large part managed to tame the water cycle – capturing water so that we can have it when we need it, rather than waiting for it to fall from the sky – is a key element of the human domestication of our planet and surely one of our greatest successes as a species.

Indeed, one of today's most urgent challenges is to bring these advantages – of clean and plentiful water – to the entire world's population. Anyone who cares about children should care about clean water, for every day 6,000 children (mostly under the age of five) die from preventable diarrhoeal diseases linked to dirty water and bad sanitation, and many more from malaria and other water-related diseases. Globally, 1 billion people lack access to an improved water supply, and 2.6 billion lack access to improved sanitation. A pledge to halve by 2015 'the proportion of people without sustainable access to safe drinking water' is one of the most important of the UN's Millennium Development Goals, and although globally the target is likely to be met, one of the regions that is still falling behind is sub-Saharan Africa, where most water and disease-related deaths of young children occur.

But hardly any progress has been made in meeting the associated Millennium Development Goal on sanitation, and current projections suggest that two thirds of the world's population will still not be connected to public sewerage systems even by 2030: a grim indictment of our current lack of commitment in this vital area. For those more moved by economic logic than by thousands of daily avoidable childhood deaths, the benefits of investment are still overwhelming: for each $1 invested in improving water supplies the World Health Organisation estimates annual returns of between $3 and $34 – rates most commercial investors can only dream of.[8]

Freshwater is vital not just for drinking but also for food production. The increasing use of irrigation in agriculture was one of the most important benefits of the Green Revolution, which began in the 1960s and – by bringing high-yielding crop varieties, fertilisers and pesticides to farmers in the developing world – meant that food prices actually fell in real terms throughout the period of rapid population growth between 1960 and 2000. Irrigation allows food to be grown in areas without sufficient natural rainfall, and as a result can be highly productive as levels of sunshine and temperatures in arid tropical areas also tend to be high. Today irrigated agriculture covers 275 million hectares and produces 40 per cent of humanity's food on just 20 per cent of our land.[9] Many activists I know have devoted years of their lives to campaigning against big dams because of the local people displaced or dire ecological impacts, but the truth is that much of the water we need for farmland irrigation is captured behind those dams – according to one estimate they support 12–16 per cent of current global food production.[10]

DAMMED ECOLOGY

But water, like all the planet's major renewable resources – fisheries, soils and forests – is fundamentally limited in terms of how much of it can be sustainably used at any one time by humans. Water differs from fisheries or forests in that it can never be destroyed and will go on circulating for ever with or without humans. But the amount of water flowing through the Earth's rivers at any one time is also a fixed quantity, which we can do little to alter. Freshwater is the lifeblood of the biosphere, and rivers the blue arteries through which this life force circulates. Use too much of it, and we risk causing severe and irreversible ecological changes, just as in the other planetary boundary areas. All living organisms need water just as they need food; as our hold on the planet's land surface has deprived other species of food, so our dominion over the hydrological cycle has increasingly deprived them of water.

Rivers are vital ecological zones: they provide habitat for plants and animals within a myriad of different ecosystems. Wetland vegetation purifies water, for example, and large wetland areas can even help generate rainfall much further afield by increasing water evaporation.[11] Rivers deliver sediments and nutrients to delta areas and coastal fisheries, whilst also absorbing pollutants like nitrogen and phosphorus, breaking down waste and delivering clean water over a far wider scale than any human sewage works. Freshwater ecosystems also harbour biodiversity far out of proportion to their area: river, lake and wetland habitats cover less than 1 per cent of the world's surface, yet provide a home for over a quarter of all known vertebrates – more than 126,000 known animal species in total, and about 2,600 aquatic plants.[12]

In Botswana's Okavango Delta, for instance, the world's largest inland delta supports an incredible diversity of animal and plant life where waters draining from the Angolan highlands spill out into the otherwise arid Kalahari Desert. These seasonal floodplains are home to buffalos, elephants, hippos, lions, wild dogs and the beautiful sitatunga antelope, which together share the fertile marshes with 1,300 types of plant and more than 400 species of bird.[13] Many animals are only occasional users of freshwater habitats, but are highly dependent on them nevertheless: in Asia land animals using swamp or riparian areas include the orang-utan, Javan rhinoceros, Asian elephants, tapirs and proboscis monkeys. Sadly, all these animals are today endangered, in large part because of habitat loss.

Globally the extinction rate for freshwater species is five times higher than for terrestrial species, and according to the latest IUCN Red List survey 37 per cent of assessed species are counted as threatened.[14] In general, because of dam-building and other human alterations to water flows, freshwater ecosystems are some of the most threatened on Earth. In the United States, originally blessed with some of the highest levels of freshwater biodiversity in the world, a shocking 123 species have gone extinct since 1900. The decline continues: 69 per cent of the 300 species of freshwater mussels are endangered or extinct, as are 51 per cent of the crayfish, 37 per cent of the freshwater fish and 36 per cent of the amphibians.[15] Conflicts between human and ecological uses of water have intensified in recent years: in California the delta smelt – a small fish that was once abundant in the Sacramento–San Joaquin Delta area – is in danger of extinction because its saltmarsh habitat is disappearing and spawning adults get chewed up in the complex system of pumps used to transfer water to the state's cities and farmland. In the same area, agricultural water use now threatens the very survival of Pacific salmon, whose populations have crashed so dramatically in recent years that fisheries have been closed and several subspecies added to the Federal Endangered Species list.[16]

Around the world, many river basins are considered 'closed', in that every drop of available water is already allocated and used, and as a result little or none reaches the sea. Perhaps the most famous closed basin is China's Yellow River, which was so heavily used by the late 1990s that the river failed to reach the sea for almost the entire year.[17] Water managers at one point recorded 687 kilometres of the river's length with no flow at all. In the Middle East, so much of the River Jordan's flow is diverted into Israel's National Water Carrier that little if any reaches the Dead Sea, whose level is now falling precipitously.[18]

Dams and reservoirs tend to iron out the peaks and troughs of high- and low-season river flows, yet many species are adapted to respond to these seasonal signals. For example, the Missouri River in the United States once experienced annual snowmelt floods, when warmer spring temperatures thawed snowpack in the Rockies and released a pulse of water into the river. These floods distributed seeds and nutrients and allowed fish to move onto the floodplain to feed. In the later summer, low-water periods would expose sandbars that were crucial for nesting birds, leave large shallow areas and braided channels essential for spawning sturgeon, and expose extensive mudflats where migrating birds would forage for invertebrates. Today most of the Missouri's flow is controlled by six large federal dams operated by the US Army Corps of Engineers. As a result of the disappearance of the natural flow regime, fish can no longer access floodplains, and crucial shallow water and bankside habitat has disappeared, whilst the flow changes that provided critical seasonal life-cycle cues to many different species no longer appear. The ecological toll has been disastrous: of 67 native fish species living in the river, 51 are now classed as rare, whilst the pallid sturgeon and two bird species (the least tern and the piping plover, both dependent on vanishing sandbanks for nest sites) languish on the federal Endangered Species List.[19] Downstream from many hydroelectric dams, river heights can vary by metres in just a few minutes as vast quantities of water are released to generate electricity – leaving animals and plants alternately swept away or stranded high and dry on the riverbank.

Another key ecological water characteristic is temperature. The Colorado River in the US and Mexico was changed for ever by the construction of the Glen Canyon Dam in 1963, after which natural seasonal changes disappeared and water was released from 60 metres down under the newly created Lake Powell at a uniform 9°C all year round. In many rivers thermal regimes are as important as flow regimes in providing seasonal life-cycle cues to various species, provoking fish to spawn as water temperatures rise and insects to emerge from pupa. Whilst water extraction leaves less water in the river's main channel – meaning the remainder tends to heat up more quickly – water released from thermally stratified large reservoirs like Lake Powell tends to be much colder than undammed flows. Large distances are needed to mitigate this thermal pollution: estimates suggest that for the Colorado's water to recover its natural temperature downstream of the Glen Canyon Dam would require 930 kilometres of flow – impossible anyway because of other downstream dams.[20]

Water trapped behind dams also loses most of its sediment load, so suspended sediment is no longer available to nourish downstream deltas, floodplains and estuaries. The Colorado River was so named because the word means 'coloured' in Spanish, in reference to the huge amounts of red sediment the river once transported. But since the closure of the Glen Canyon Dam this once-turgid river has flowed crystal-clear, nearly eliminating native fish species that were adapted to feeding in muddy waters without eyesight. (Only three out of eight native fish species that were abundant before 1963 are still common today – the rest are extinct or nearly so.) Worldwide, reservoirs trap about a third of the sediment that would otherwise flow down rivers, and more than 90 per cent in heavily dammed systems like the Nile, the Colorado, the Rio Grande and the Volga.[21]

Where river deltas are denied a constant influx of sediment, they lose the battle against the ocean, and begin – slowly but irrevocably – to sink. This is bad news for the megacities built on deltaic plains like Shanghai on the Yangtze delta, Calcutta and Dhaka on the

Ganges–Brahmaputra, Lagos on the Niger and Ho Chi Minh City on the Mekong delta. As these cities approach or even drop below the level of the rising seas, the risk of flooding intensifies: 10 million people a year currently suffer from storm surges, mostly in Asian deltas. Direct hits by hurricanes can be catastrophic, as New Orleans discovered during the Hurricane Katrina disaster in 2005. In the Burmese Irrawaddy delta a storm surge is estimated to have killed upwards of 100,000 people during Cyclone Nargis in 2008. Deltas where rivers no longer reach the sea get no sediment deposition at all, like the rapidly disappearing and ecologically sensitive Colorado delta in Mexico.

Given all of these ecological impacts, and the natural upper limit set by the amount of freshwater flowing through the world's rivers at any one time, humanity has little option but to recognise that water must be shared between our species and the rest of the biosphere if major and accelerating impacts on the Earth system are to be avoided over the long term. The question is where this level should be set, and the planetary boundaries expert group suggests a quantified boundary, expressed in terms of the human consumptive use of freshwater, of 4,000 cubic kilometres per year. Current human consumptive water use (meaning water used, evaporated or incorporated into a crop, so not available for direct reuse) is estimated at around 2,600 km^3.[22] So on a planetary level, humanity is still within the confines of what can be considered sustainable, despite all the specific environmental impacts listed above.

As with the land use boundary, this is qualified good news. It means that there is still some room for expansion of human consumption before we run up against hard-wired limits in the Earth system in some areas. Whilst on climate, biodiversity loss and nitrogen we are well over the levels set for sustainable long-term use, we can still tap some rivers to bring freshwater to those who currently lack it and to grow more food for a still-expanding population. But where this water is taken from matters hugely: whilst greenhouse gases have the same impact wherever they are around the planet, water removed from an arid region like the Middle East will be much more damaging than the same amount captured from a river in well-watered northern Canada. This planetary boundary is therefore perhaps best thought of as an aggregate total of many smaller boundaries operating at the scale of single river systems in different continents and regions, rather than as a single global number in and of itself.

WET AND WILD

The most radical solution for restoring rivers to ecological health is simply to tear down unnecessary dams, revitalising the natural hydrology and allowing migrating fish to return. Anywhere this can be done it should be, and much evidence suggests that a substantial fraction of currently existing dams could and should simply be removed. The United States is at the centre of a new movement towards freshwater ecological restoration: in one of the most unsung environmental successes anywhere in the world, over 800 dams have been demolished in recent years and thousands more could follow.[23] Rivers can bounce back to health remarkably quickly once concrete impediments are removed: after the destruction of the Edwards Dam on Maine's Kennebec River, 2 million alewife fish returned within a year to spawn, and American shad, striped bass, Atlantic salmon and sturgeon all followed. In September 2011 the tallest dam yet – the 210-foot-high Glines Canyon Dam – is slated for removal, aimed at restoring over 70 miles of salmon and steelhead habitat. The following month, and with the support of the Yakama Indian Nation, the Condit Dam on the White Salmon River will also be breached.[24] Given the massive decline in salmon populations seen recently in the American northwest, the restoration of natural flows to rivers in the region is an urgent priority for fishermen, Indian first nations and conservationists alike.

Where outright removal of dams is not a pragmatic option, there are various technical interventions that can be made to mitigate their ecological impacts. Reservoir intakes can be sited at different levels to allow operators to release the right temperature water from different depths in the lake in different seasons. Water managers can also pay more heed to maintaining 'ecological flows', mimicking a river's natural flow regime whilst simultaneously allowing for competing hydroelectric production and irrigation demands – as the US Army Corps of Engineers is now beginning to do in the Missouri. For the Colorado River through the Grand Canyon, experimental human-engineered floods have re-established habitat and rebuilt sandbanks. Also in the US, the Army Corps of Engineers and the Nature Conservancy have embarked on a nationwide partnership to find ways for dam operators to mimic seasonal cycles to restore riverine ecosystems. This Sustainable Rivers Project is currently working on 11 rivers with 26 dams flowing through 13 states.[25] Even so, that is still only a tiny proportion of the rivers being affected by dams, and only in one country. Worldwide, we need a revolution in the way that water managers control rivers, in order to arrest the decline in freshwater ecosystems by maintaining ecological flows to the furthest extent possible.

In some parts of the world the removal of water from rivers for agriculture has caused such havoc that radical and costly measures will be needed to reverse the damage. Probably the worst water-related catastrophe anywhere in the world afflicts Central Asia, where the Aral Sea – once the world's fourth-largest inland lake – all but disappeared after Soviet managers in the 1960s and 1970s diverted its two main feeder rivers for cotton production. What was once a thriving wetland ecosystem, supporting large-scale commercial fishing as well as numerous waterbirds and other species on river deltas, has been turned into an ecological disaster area. Whole former port towns have been left stranded tens of kilometres from the water line, and two of the remaining three remnant portions of the former Aral Sea are so salty that they support no fish at all. The whole area is raked by toxic dust storms, countless thousands have lost their livelihoods and the region's wildlife has been decimated.[26]

Almost as appalling as the scale of the original damage has been the weakness of the response. Only the small northern part of the basin has been partially protected to restore a semblance of its original water levels, and only at the cost of removing even more water from the south with a dyke. Instead of emptying into the Aral Sea, the southern feeder river, the Amu Darya, now disappears into the desert – all its water is still captured to grow Uzbekistan's precious cotton, and the situation with the northern Syr Darya river is little better. Even though the Soviet Union has now disappeared, its successor states have equally entrenched and conservative interests preventing even a fraction of the original flows of the Aral's rivers ever being restored. Perhaps the most unpleasant of these political forces is Uzbekistan's cotton industry, which is unproductive, toxic and substantially dependent on enforced child labour. Cotton benefits only the country's despotic government, and I would urge Western consumers and companies to boycott the industry completely.[27]

The Aral Sea could still be substantially restored if the political will were mobilised to make this happen. Given an ambitious effort to restore the original flows of the two rivers, experts have estimated that the entire Aral basin could be refilled in less than forty years.[28] The effort would require the large-scale abandonment of thirsty cotton crops, and free markets to allow farmers to choose what they want to plant – instead of being dictated to by the Uzbek government. As well as a reduction in irrigated land overall by more than half, major investments need to be made to increase the efficiency of water infrastructure so that less of it is wasted via leaching and evaporation from canals. Farmers need to be provided with alternative livelihoods so that they are not even further disadvantaged as the focus moves away from agriculture in what is after all a very arid area. Given that what happened to the Aral represents the worst regional environmental disaster in the world, it is in the interests of the entire international community to restore as much of the damage as possible – and to pay for the necessary process of social transition that this implies.

Much of the Aral Sea's vanished river inflow is captured for upstream hydroelectricity, demonstrating the tradeoffs inherent in meeting the water planetary boundary. One of these is with the climate change boundary, for hydroelectric dams produce by far the largest fraction of the world's renewable electricity – currently about 16 per cent of the global total.[29] China's controversial Three Gorges Dam displaced 1.3 million people[30] (submerging an astonishing 13 cities, 140 towns and 1,350 villages) and further disrupted the ecology of the already highly impacted Yangtze basin, but produces as much power as 20 big coal plants, avoiding the emission of perhaps 100 million tonnes of carbon dioxide every year. Fifty more dams are planned on the Yangtze by the Chinese government, which wants to double hydropower's contribution to the country's electricity supply by 2020 to a hefty 300 gigawatts. Given that China is already going flat-out for nuclear and wind – both better options than hydro in terms of overall ecological impacts – this is a serious conundrum, and one that is repeated globally in many rapidly growing developing countries like India. As always a balance will have to be found, based not just on economics but also on the environmental costs of competing forms of power generation – and in each case all the planetary boundaries must be considered.

One type of hydroelectric generation that certainly needs expanding is pumped storage. In a pumped storage plant, water is pumped uphill into a higher reservoir when electricity is cheap and plentiful on the grid (such as at night) and then released during periods of high demand. In many countries pumped storage systems are an important way to balance loads on the electrical grid, as they can respond very quickly to additional demands for massive amounts of power. The largest system in Europe, at Dinorwig in Wales, can go from 0 to 1.3 gigawatts of generation in 12 seconds (and maintain this rate for several hours), and the turbines are reversible as pumps to return the water to the higher reservoir once the peak of demand has passed.[31] Pumped storage systems have minimal ecological impacts if they are closed-loop, off-river systems, where the water is used over and over again. Because these systems essentially act like enormous batteries, they represent the only existing grid-scale option for storing power, and are likely to be an essential part of the energy mix as renewable and nuclear sources form an ever-larger part of the grid in more and more countries.

Also interacting with the climate change boundary is the option of desalination, which allows farms and cities in arid areas to use water from the sea by extracting the salt to yield freshwater. If Israel could use more desalinated water, then perhaps some of the flows of the River Jordan could be restored and the gradual drying up of the Dead Sea avoided. In desert countries like Saudi Arabia and small island states that have no rivers at all, desalination is the only option to support current human populations. But there is a price: desalination is a very energy-intensive process, and where fossil fuels are used contributes substantially to greenhouse gas emissions. In Saudi Arabia, currently the world's largest producer of desalinated water – on which it depends for 70 per cent of its drinking water – the largest plant at Shoiba sits next to an oil-fired power plant, which supplies the required heat and energy.[32] But desalination could be very compatible with intermittent renewable power sources like solar and wind, because freshwater – unlike electricity – can be easily stored, so only needs to be produced when the wind is blowing or the sun shining. Care however needs to be taken to ensure that the concentrated brine that is left over from desalination does not damage the marine environment when discharged into the ocean. Overall, desalination powered by renewables can likely make a major contribution to helping arid countries take more water from the sea, and thereby extract less from ecologically sensitive and overused river basins.

Once again, global warming will change everything. Although precipitation levels globally are expected to rise, the changing climate will alter the distribution of the planet's water. Unfortunately most of the additional rainfall is expected in higher latitudes, whilst dry subtropical areas like the Mediterranean,[33] southern Africa, Central America and Australia can expect to get dryer still.[34] Here existing dense populations may make additional water storage or desalination unavoidable. A changing climate will also change the seasonality of runoff: in the western United States, for example, earlier snowmelt and more winter rainfall (rather than snow) in a warmer world could advance the date of peak runoff by as much as two months.[35] In high mountain areas like the Andes and Himalayas, melting glaciers will change the hydrology of downstream rivers, generally meaning less river flow in the dry season. For these regions new reservoirs are about the only conceivable adaptation option: natural frozen stores of water must be replaced with human-engineered liquid ones.

MAKING WATER PAY

Two of the most effective remedies for water scarcity are policies more usually associated with the political right than with the left: increasing global trade and, most controversially of all, privatisation. For years now it has become an accepted mantra that we should all eat locally produced food in order to reduce the volume of global trade and help the climate. There are many reasons why this is a good idea: I am a devotee of my local farmers' market because I like to support growers in my area and get to know the people who are producing my family's food. But environmentally the issues are less clear-cut. Although some greenhouse gases are involved in the shipping of bulk commodities like wheat and beef, in water-use terms it makes sense for most food to be produced in well-watered areas with high rainfall rather than in arid regions where irrigation can devastate the local ecology – as it has in Central Asia's Aral Sea. The water used in producing different crops and foods is rated as 'virtual water', and its trade provides a little-known but enormous global environmental boon.

In Egypt, for example, each kilogram of maize imported from Europe saves this desert country half a cubic metre of water.[36] Wheat imports from abroad in total therefore save Egypt 5.8 billion cubic metres of water each year, equivalent to 10 per cent of its national water use. 'Going local' to produce bread for Cairo would require more water to be extracted from the Nile and pumped to farmland reclaimed from the desert, where high evaporation would mean heavy losses and salination of the land. Saudi Arabia has sensibly decided to phase out growing wheat domestically using fossil water pumped unsustainably from underground and rely on imports from 2012.[37] Nearby United Arab Emirates is likewise mulling over abandoning most domestic agriculture, even fodder for its beloved camels.

Studies of the amount of water saved globally per year through the virtual water trade in food amount to some 455 cubic kilometres,[38] a substantial saving given that only 2,000 cubic kilometres remain to be used before we find ourselves the wrong side of the proposed water planetary boundary. Producing food where the maximum water efficiency can be achieved, and where water is most plentiful, is a natural extension of the idea of comparative advantage, a basic concept in economics that was first proposed by David Ricardo in the early nineteenth century. The idea does have relevance domestically as well as internationally: instead of trying to divert enormous quantities of water from the wetter south to the arid north, China could instead explicitly formulate a policy to focus on transferring virtual rather than real water, saving money and promoting environmental efficiency in the process. Real water is bulky, heavy and leaky, so transferring it large distances is both costly and inefficient, as well as damaging to the hydrology of source regions.

Unfortunately, global water savings through trade are currently largely accidental, and few governments make efforts to promote international trade in food on this basis. Some countries are also highly anomalous: well-watered Switzerland, for instance, imports a huge amount of virtual water in food, whilst arid Pakistan and Australia both export vast quantities.[39] Given that financial and economic arguments invariably carry more weight with policymakers and governments than environmental ones, probably the only viable long-term solution will be for agriculture to be forced to pay properly for the water it uses. Currently farmers around the world are either given water for free or allocated highly subsidised reserved amounts, giving them little incentive for conservation or efficiency, and meaning that the final price of food crops does not reflect the value or scarcity of the water used in producing them.

There is a simple solution to this: privatisation. The provision of water must be deregulated and privatised; taken out of the inefficient and often corrupt hands of the state, and handed instead to the private sector. I would also suggest that the World Trade Organisation focus more on eliminating water subsidies to farmers to free up trade barriers and make the international trade in virtual water more efficient. I am aware that this proposal will be controversial, and that water privatisation is typically opposed by NGOs and social movements on the grounds that it will disadvantage the poor. In some cases outright civil conflict has resulted from privatisation proposals: in Cochabamba, Bolivia, in 2000 demonstrators were fired on by police during marches against water privatisation, resulting in a state of emergency and several deaths.

But opposition to privatisation makes the implicit assumption that the public sector is doing a good job – a notion that fails to stand up to empirical analysis. In fact, most state provision of water is rather inefficient and the public companies providing the service are frequently starved of investment due to pressure on government budgets. Many international studies now show that privatisation tends to increase the number of piped household connections, benefiting the poor, who no longer have to buy bottled water at a premium or rely on dirty wells or rivers.[40] Privatisation of water in Argentina has even been linked with a decrease in child mortality, confounding the expectations of leftist social campaigners.[41] The reason is simple: private companies seek to make a profit, so in a competitive market they tend to increase the efficiency of their service to customers so as not to lose earnings unnecessarily.

Just as with climate change, exposure to free markets alone will not solve the world's water problems. We all have a part to play in reducing water wastage at the household level. More water-efficient toilets, for example, can be useful in reducing the absurdity of clean drinking water being used to flush away human sewage, whilst as an individual you can take showers rather than baths and turn off the tap when brushing your teeth. In the garden, ditch the hosepipe and collect rainwater instead – especially if you are making a wildlife pond, where the chlorine in most piped water will damage water life. Give up bought bottled water too: not only does this save on plastic packaging and transport, but piped water is also less carbon-intensive than bottled mineral water, and may even be safer too. For households, water meters are a good way of giving people a financial incentive to conserve, and should be retrofitted and made standard wherever possible.

Governments must get involved in this effort. Putting a price on water that truly reflects its ecological scarcity needs the agreement of world governments just as does putting an international price on carbon. The point of a carbon price is to correct the market failure by which fossil-fuels producers and consumers pay nothing for the damage greenhouse gases cause to the climate, and hopefully thereby to eliminate carbon emissions. Aiming for zero water however is not an option, because water is a biological necessity both for ourselves and for agricultural production of any sort. Nor is zero water use necessary, given that water is an indestructible and renewable resource. But the efficiency of our use of water can and should increase, so that we use less of it at any one time. Israel has shown the way forward by recycling household water for use in agriculture on a massive scale: more than 80 per cent of waste water is recycled for use in producing food.[42]

It might also be possible for the concept of carbon markets to be extended into the realm of water. Just as carbon credits represent a virtual trade in greenhouse gas permits (no one actually moves tonnes of real carbon dioxide around when they buy or sell, although they may change the site of its emission), so the virtual trade in water might be enhanced by the trading of water credits so that arid regions that save on using water in farming or industry get a financial bonus for helping the world meet the water boundary. Getting cap and trade to work for water will be complex, however, because water is not essentially fungible like CO_2, which has the same global warming potential wherever it is released. As I stated earlier, a cubic metre of water in the Jordan is much more valuable than a cubic metre in the Hudson, so any pricing and trading system must somehow reflect this scarcity differential. This apples and oranges problem does not have to be a deal-breaker, however – in the international carbon markets a whole basket of different greenhouse gases (including methane, nitrous oxide and HFCs) are traded interchangeably despite them having very different effects on the atmosphere over different time-scales. Markets are human instruments, and can be targeted to achieve any environmental objective if cleverly designed with that end in mind.

Just as the localisation of agriculture may be ruled out by water constraints, industrialised farming and the Green Revolution have likely helped humanity with regard to this planetary boundary. In France, for example, wheat yields per hectare tripled between 1960 and 2000 as new technologies dramatically increased productivity.[43]

In other words, three times as much wheat was being produced per unit of both land and water by the turn of the century as forty years earlier. Wheat in France is mostly rainfed, but the same logic applies in any place where increasing yields outstripped any increase in water use. For the whole European Union, agricultural productivity gains over this period brought a saving of 1,800 litres per person per day – a bigger saving than any other measure that could have been taken, and one that (in water terms anyway) happened entirely by accident. Similar patterns applied both in America and globally. In all probability, therefore, the Green Revolution stopped us already breaching the water planetary boundary a quarter of a century ago.

CHAPTER SEVEN

The Toxics Boundary

There is perhaps no better illustration of the domestication of our world than how rapidly we are turning it into plastic. Plastics are everywhere. In the deepest ocean trenches, plastic bottles are silently accumulating in thick drifts. Submersibles have photographed plastic bags suspended eerily above the seafloor a mile below the Arctic ice cap. In other sea areas 10,000 man-made items have been found scattered over a single hectare of the ocean bottom.[1] Plastic from the ocean is breaking down and accumulating on beaches: one sand sample gathered from a beach near Plymouth, England, contained 10 per cent plastic particles by weight.

In remote islands, seabird populations are being devastated by plastic refuse. On Midway atoll in the Pacific, albatrosses are estimated to feed their chicks a combined 5 tonnes of plastic each year, and 200,000 out of the half-million chicks die each year as a result because of dehydration, perforated stomachs or starvation. Wildlife rangers typically find cigarette lighters, toothbrushes, syringes, toy soldiers, Lego and all manner of other items in dead chicks' stomachs. 'The atoll is littered with decomposing remains, grisly wreaths of feathers and bone surrounding colourful piles of bottle caps, plastic dinosaurs, checkers, highlighter pens, perfume bottles, fishing line and small Styrofoam balls,' reported one visiting journalist.[2]

The problem with plastic, as with toxic wastes generally, is that the natural world has no way of biologically decomposing artificially manufactured polymers. Humans have now devised countless thousands of novel substances, never before seen on Earth, and released them into the natural environment. Many appear perfectly benign at first pass – but turn out to be very different when they enter the food chain, whether on land or in the sea. Some of the toxins we produce are naturally occurring, like mercury, but human activities mobilise or concentrate them in ways that are potentially devastating to other species and, ultimately, ourselves.

To a large extent the modern environmental movement was founded out of a concern about toxins. Rachel Carson's seminal book *Silent Spring* brought attention to the wanton dispersal of agricultural pesticides like DDT, which she showed were having a damaging effect on non-target species higher up the food chain. Many Green groups including Greenpeace and WWF have devoted decades both to researching the impacts of toxic chemicals, and campaigning for their proper regulation. I have no argument with the Greens here: the movement's work has by and large been pragmatic and effective. It has also shown a willingness to compromise, and to face up to real-world conflicts. DDT, for instance, is important in controlling malaria, and international regulations banning its use in agriculture allow necessary applications to control mosquito populations – exceptions that are sensibly supported by Green groups.

GENDER-BENDERS AND VULTURE KILLERS

There is every reason to be cautious about the impacts of toxic pollution. We know that man-made chemicals can have serious and unanticipated impacts on wildlife even at very low doses. In India the numbers of oriental white-backed vultures fell by 99.9 per cent between 1992 and 2007, due to a mystery killer that was eventually traced to residues of the anti-inflammatory drug diclofenac within the rotting corpses of livestock that were picked apart by the vultures. Only a thousandth of the original population of oriental white-backed vultures still remained by the late 2000s – a rate of decline that was, as zoologists pointed out, faster than that of any other wild bird, including the dodo.[3] And the decline continues – only the manufacture, not the use, of diclofenac is banned, and the drug remains popular with farmers. The three main Indian vulture species are now classified as critically endangered, their extinction staved off only by emergency captive breeding programmes and vulture 'restaurants' where clean carcasses provide food for the remaining few birds.

Some of the most troubling chemicals are the so-called 'gender-benders', endocrine-disrupting substances whose effect is to mimic or block the action of natural biological hormones in an animal's body. How exactly this effect takes place is still being studied, but some of the most notorious pollutants – DDT, lindane, organophosphates, PCBs – are thought to affect animal sexual biology, altering the reproductive capabilities of everything from bulls to alligators.[4] The phenomenon was first noticed in England on the River Thames, where 30 years ago fishermen noticed that there was something wrong with the roach they were catching downstream from a waste-water treatment works. The fish were neither male nor female but intersex: the male gonads contained both testicular and ovarian tissue.

Scientists suggested that 'environmental oestrogens' could somehow be to blame, by altering hormonal balances in the bodies of the fish and thereby affecting their sexual development.[5] And sure enough, studies of the water composition found synthetic oestrogen from the contraceptive pill (excreted by women in urine and not removed by sewage treatment) contaminating the river, along with horse oestrogen used in human hormone-replacement therapy and alkylphenolic chemicals from cleaning agents and paints. In fact, aquatic animals seem to be particularly at risk from gender-bending chemicals. In the late 1980s it was discovered that tributyltin (TBT), a compound used in anti-fouling paints on ships' hulls, was turning dog whelks 'imposex' (male into semi-female and vice-versa) all along the coast of the English Channel.[6] Some female whelks were even found to be growing penises, blocking the release of eggs and rendering large numbers sterile.[7]

Another worrying property of man-made toxins is their capacity to accumulate in the food chain. Indeed, this propensity to 'bioaccumulate' was recognised early on for DDT, which was affecting animals at the top of food webs apparently out of all proportion to their direct exposure. Many chemicals are also extremely mobile, and can affect species a long distance away from their original site of release. Large quantities of the herbicide atrazine, commonly used on corn in the

United States, fall out of the sky in rain or snow – and measurable quantities have been observed in precipitation in otherwise pristine national parks.[8]

Due to the dynamics of weather systems and long-range atmospheric transport, many of these toxins accumulate in the polar regions, particularly the Arctic. In 2009 scientists measuring contamination levels in the eggs of Norwegian and Russian ivory gulls discovered levels of pollutants – including DDT, mercury, PCBs and brominated flame retardants – at unprecedentedly high concentrations, enough to potentially affect the reproductive success of the birds.[9] The ivory gull population in the entire Arctic is thought to be only around 10,000, and studies in the Canadian Arctic show an 80 per cent fall since the 1980s. In response to this decline, the gull has now been reclassified on the IUCN Red List from 'least concern' to 'near threatened'.[10]

Although PCBs were largely banned in producing countries by the end of the 1970s as their toxic potential became better known, these chemicals continue to accumulate in the Arctic. A recent survey of ponds near seabird colonies found high levels of PCBs in water nearest the bird-nesting and roosting sites, showing that toxic chemicals are still being transported to the far north via the marine food chain.[11] Also at high levels was the poisonous heavy metal cadmium, which was present at 16 times background concentrations in pond beds near the seabird colonies – suggesting that it too bioaccumulates through the food chain.

Being at the apex of the food web, human beings are especially at risk. In the late 1980s researchers in Arctic Canada discovered that Inuit women had breast-milk concentrations of PCBs up to ten times those of women further south, thanks to the Inuit 'country food' diet of seals, walrus and beluga whales, all of which accumulate toxins in their fat deposits.[12] More recent surveys have confirmed that Inuit people are particularly exposed to chlordane, toxaphene and PCBs, with mercury levels high enough to pose a risk to unborn children.[13] In other parts of the world people are warned not to eat fatty or carnivorous fish because of the dangers of bioaccumulated toxins.

GOOD FOR GONADS?

Their possible effects on humans make environmental toxins one of the more successful campaigns mounted by Greens over recent decades. No one likes to be poisoned, and the fact that many of these chemicals are proven carcinogens increases public support for their elimination. Equally worrying is their sheer ubiquity: out of 3,000 or so pharmaceutical products in everyday use, 100 have already been found in rivers and watercourses. Considerable controversy now surrounds the possible effects of phthalates and bisphenol-A, especially given that their use in baby bottles may expose newborn children to low levels of endocrine-disrupting chemicals. Both are widely used in the modern world, including in vinyl floors, cling film, cosmetics, medical products and toys.

Even treated drinking water commonly contains traceable amounts of potentially toxic chemicals. One 2009 study of pharmaceuticals and endocrine-disrupting chemicals in US drinking water looked at 19 drinking water treatment facilities serving 28 million people, testing for contamination by 51 different compounds. The researchers identified 34 of them in at least one sample of drinking water. Most common was the omnipresent herbicide atrazine, but also spotted were diazepam (Valium), naproxen, risperidone (an anti-psychotic) and fluoxetine, though none at levels deemed unsafe.[14] It is a truism to say that the poison is the dose, and it is important for context to bear in mind that these contaminants may be having no effect at all at the levels identified, particularly as some of them are drugs used routinely in medicine with well-understood and minimal side effects. The levels were also much lower than commonly found residues of water disinfection like chlorine.

Indeed, there is a great deal of media mythology that has grown up around the issue of human health and fertility in regard to potentially toxic chemicals. One particularly stubborn myth – almost universally believed in my experience – is that sperm counts are falling around the world, and that this threatens a worldwide crisis in male fertility. Although certainly some studies over the last twenty years have shown declining sperm counts, many more have shown no change at all. There are also great differences in sperm counts between different men according to lifestyle factors and geographical location: counts in New York have been found to be double those in California for no obvious reason.[15] Smoking is known to reduce sperm counts, as is obesity. Counts even vary in individuals, due to the recent temperature of the testicles and the time gone by since the last ejaculation.

There are also methodological obstacles that make this particular branch of medical science rather challenging. Because volunteering a sperm sample requires the subject to masturbate, it is virtually impossible for emotional and cultural reasons to get enough people from a broad enough random group to participate to make a study generally meaningful. 'If collection of semen samples were as straightforward as obtaining blood samples, the nature of semen quality changes over time (if any) would have been determined decisively decades ago,' one expert complains.[16] Instead, studies tend to concentrate on groups of men who are producing samples for other reasons, such as donating to sperm banks, or are required to take tests because their partners cannot conceive – potentially biasing the results of studies depending on them. Obviously, the fertility clinic is not the best place to get a meaningful sample of sperm counts in the wider population.

Another myth is that other indicators of male fertility are also in worrying decline. These include hypospadias, a relatively common birth defect in males where the opening of the urethra is found not at the tip of the glans but along the underside of the shaft of the penis or even – in severe cases – in the scrotum, as well as cryptorchidism (the failure of one or both testes to descend into the scrotum in very young boys) and testicular cancer. Although some studies have shown hypospadias increasing in the US and Europe over the last quarter-century,[17] a great many more – using newer data – have shown no change at all,[18] and it seems reasonable to conclude that there is no strong evidence implicating environmental chemicals in any change in male fertility in recent decades.

The same conclusion applies for women, too: a 2004 meta-study, for example, did not find any correlation between DDT exposure and breast cancer – both varied enormously among the sampled populations, but there was no relationship between the two.[19] Another study in Long Island, New York, looked at breast cancer and pesticides, DDT and PCBs in several hundred women, but found no statistical link between those who suffered from cancer and those whose bodies contained the highest levels of putative toxins.[20] None of this is to say that at more potent doses many of these compounds may not be toxic or carcinogenic, nor that they should not be strictly regulated in order to reduce the risk of health impacts. But at the very low levels most of us experience, the prevailing scientific wisdom is that they probably do us no harm at all.[21]

ENIGMATIC DECLINES

In the natural world, things are no less mysterious. For nearly a decade now researchers working in remote locations from Central America to Papua New Guinea have been finding dead or dying frogs, often strewn gruesomely around in scenes reminiscent of a field of battle. Sometimes only the skeletons remain, or just an empty pond, once teeming with amphibian life but now with scarcely a tadpole to be seen. Even where frog populations have not been utterly wiped out, many have been suffering an epidemic of strange deformities.[22] Adults show up with two extra legs, misshapen eyes or unpleasant skin lesions. Recent extinctions of entire species have been reported from Australia, Costa Rica, Honduras, Israel, Kenya and Mexico. Many more are reported as 'missing in action', not sighted in the wild for several years and presumed extinct. The phenomenon is known as 'enigmatic decline', and environmental contaminants have been suggested as being at least partly to blame.

Laboratory studies have indicated impacts at very low concentrations of toxins, with worrying implications. For the herbicide atrazine, researchers at the University of California in Berkeley in 2009 performed an experiment where male African clawed frogs were exposed to the chemical in water – at extremely low levels of 2.5 parts per billion, classified by the US Environmental Protection Agency as safe for human drinking.[23] Several of the atrazine-exposed male frogs were 'chemically castrated' or 'completely feminized' as adults, according to the subsequent report.[24] This all sounds scary, even definitive, but the scientific rigour of the study has been questioned: proper science demands that a reported conclusion can be independently replicated or repeated, but other studies on atrazine's impact on the same species of frog have found no impact on growth or sexual development,[25] and the idea remains extremely controversial.[26]

For amphibians in general, most scientists now believe that a fungal infection is the leading cause of declines and extinctions even in otherwise undamaged habitats, rejecting other candidate threats such as climate change and environmental toxins. This fungal pathogen, known as chytridiomycosis, now 'poses the greatest threat to biodiversity of any known disease', according to a recent report in a US science journal.[27] It certainly seems to be extraordinarily virulent: in May 2010 one group of researchers working in California's Sierra Nevada mountains reported that a chytrid outbreak sweeping through the native amphibian population reduced the number of adult frogs in one valley site from 1,680 to a mere 22, with total extinctions noted elsewhere.[28] No one knows how the fungus spreads: one theory posits that it may even be coming in on the boots of visiting amphibian researchers. If this is true, the usual conclusion that 'more research is needed' may not in fact be justified.

NO BOUNDARY, NO PROBLEM?

There is no quantified planetary boundary proposed for toxic pollution. Partly this reflects the poor state of our knowledge about the issue. We simply do not know how many chemicals are circulating in the environment or what all their effects might be. The best-known and documented cases of chronic toxin poisoning, from India's vultures to the effects of TBT on marine molluscs, came as a surprise, whilst the real-world harmful effects of the thousands of potential toxins depends strongly on their dose and differing effects in different species. As the planetary boundaries expert group concludes: 'It is impossible to measure all possible chemicals in the environment, which makes it very difficult to define a single planetary boundary derived from the aggregated effects of tens of thousands of chemicals.'

Ultimately, 'a chemical pollution boundary may require setting a range of sub-boundaries based on the effects of many individual chemicals combined with identifying specific effects on sensitive organisms,' the expert group suggests.

The toxics issue actually provides a good case study for the sensible use of regulation under conditions of profound scientific uncertainty. Most of those chemicals that are known beyond reasonable doubt to have toxic effects on both humans and wildlife are already strongly regulated at the international level. The Stockholm Convention on Persistent Organic Pollutants was adopted in 2001 and came into force in 2004, banning the production and use of some of the most damaging endocrine-disrupting and long-lived chemical pollutants. The initial 'dirty dozen' covered by the Convention included nine pesticides (aldrin, chlordane, DDT, dieldrin, endrin, heptachlor, hexachlorobenzene, mirex and toxaphene); two industrial chemicals (PCBs as well as hexachlorobenzene, also used as a pesticide); and unintentional by-products, most importantly dioxins and furans. At a Convention of the Parties to this international agreement in May 2009 in Geneva, nine other chemicals were listed for eventual elimination. These include various flame retardants, by-products of the hazardous pesticide lindane (itself due to be eliminated), another agricultural pesticide called chlordecone, and PFOS, which is found in products such as electrical and electronic parts, fire-fighting foam, photo imaging, hydraulic fluids and textiles.[29] This all represents good progress, and the agreements and efforts of campaigners and policy-makers should be applauded.

Where less is known, the approach to toxics regulation provides a reasonably good model of the precautionary principle in action. In Europe the REACH (Registration, Evaluation, Authorisation and Restriction of Chemicals) legislation, passed in 2007 and supervised by the new European Chemicals Agency, requires all companies producing more than one tonne of any novel chemical per year to safety-test and register the substance – and all chemicals sold in the EU must be covered by 2018.[30] The EU agency expects 30,000 chemicals to be tested and registered by the 2018 deadline, a number so large that many experts warn that testing facilities will be stretched to the limit as a result.[31] In April 2010 lawmakers proposed similar legislation in the United States.[32] With tens or even hundreds of thousands of different chemicals now being produced and consumed within human society, this regulatory approach is focused on trying to balance the need to avoid future disasters whilst not putting too great a strain on industry. Having largely achieved its aims, Greenpeace has wound down its chemicals campaign as a result.[33]

Even with continuing concerns about trace chemicals polluting our watercourses or leaching out of plastic baby bottles, it is indisputable that the situation in Western countries is significantly better than just a few decades ago. In England, salmon and even otters have returned to the Thames, which was once little more than a running sewer. In all developed countries, aquatic biodiversity has improved in the most polluted rivers of yesteryear, simply because sewage sludge is now properly treated, and releases of toxins by industrial enterprises are strongly regulated by national and regional environmental agencies. In rapidly industrialising countries, by contrast, horror stories continue to filter out from places like China about vast pollution incidents, and extraordinarily damaging levels of contamination. All of these arise from a failure of regulation, and can only be dealt with by national authorities having the capacity, the will and the funding to properly supervise industries and to mandate the construction of proper sewage treatment and other pollution controls. Rising levels of prosperity in the developing world will help to address this problem, for they bring with them greater levels of environmental commitment by citizens and a greater ability of governments and companies to pay for technical solutions to curb toxins emissions.

As always there are crossovers and conflicts with other planetary boundaries. The toxic heavy metal cadmium has found a new role as a major constituent of otherwise environmentally friendly solar panels, some of which use cadmium telluride to help generate electricity from the sun's rays. In April 2010, the 'green' solar industry was fighting to remain exempted from EU law governing the use and disposal of toxic substances in electronic and manufactured products.[34] The issue has split the industry, with other solar companies backing a campaign group called the Non-Toxic Solar Alliance.[35]

Clearly no business lobby should be able to circumvent toxins regulation, even if it does help to reduce carbon emissions. For as long as cadmium remains an essential constituent of some solar photovoltaic cells, steps must be taken to ensure that any leakage into the environment (for example, during a fire in a house with solar panels on the roof) is minimal, and that panels are properly recycled at the end of their lifetimes.

One area where the climate and toxics planetary boundaries clearly reinforce each other concerns the need to eliminate the use of coal as a way of generating electricity in power stations. Coal-burning stations are a major source of mercury, a potent neurotoxin that accumulates in the food chain, and other pollutants released by coal plants are equally if not more damaging. The obvious substitute for coal as a centralised form of baseload generation is nuclear, as I have made clear already, but here we come up against a central plank of conventional Green ideology. I repeat: the anti-nuclear position of many Greens does not stand up to rational, never mind scientific, examination, and the refusal by NGOs and political parties to reconsider their stance on nuclear harms both their credibility and the wider interests of the planet. I realise this is a strong and contentious statement, particularly in the light of the March 2011 Fukushima post-tsunami nuclear disaster in Japan, so I will devote the rest of this chapter to an in-depth investigation of the nuclear issue. If you are already persuaded, you have my consent to skip it. Otherwise, please read on.

NUCLEAR NIGHTMARES

'Cancers, birth defects, genetic damage, lowered immunity to diseases: these are only some of the potential effects of nuclear testing, uranium mining, radioactive waste burial and all the phases of nuclear weapons and nuclear energy production,' asserts Greenpeace International on its website, in a statement fairly typical of those made by anti-nuclear campaigners.[36] Given that we all carry radioactive elements like strontium-90 in our bones and teeth thanks to atomic weapons tests and other uses of nuclear technology since the dawn of the nuclear age in the 1950s, it does indeed seem reasonable to be concerned about possible health effects of these, as with other potentially dangerous man-made substances. So what does the science say?

First, we need to look at the sources of the radiation doses received by humans and other animals, both natural and artificial, in order to give some context to the following discussion. Background radiation permeates the environment, being continually emitted by naturally occurring radioactive elements like uranium, thorium and potassium. Radiation is also received from the sun, and cosmic radiation from the wider universe. These natural sources can vary enormously: in the UK, inhabitants of Cornwall are exposed to higher levels of radioactivity because of radon (a product of radioactively decayed uranium) emitted from granitic rocks. The UK Health Protection Agency sets an annual dose limit of 1 milliSievert (mSv) of radiation, yet annual doses in some homes from naturally occurring radon can approach 100 milliSieverts – a hundred times the prescribed safe dose.[37] The largest amounts of artificially derived radiation, on the other hand, come from medical exposures like X-rays – totalling annually about 0.4 mSv for each British person on average. In comparison, fallout from nuclear weapons tests totals 0.006 mSv, whilst contamination from civil nuclear power is lower still – even someone living outside the fence of a British nuclear power station would typically receive only 0.001 mSv, three orders of magnitude below natural background levels. Ironically enough, nuclear power stations could not be built in much of Cornwall, because the fenceline legal radiation limits would be breached due to the presence of naturally-radioactive granite.

Ionising radiation causes damage to the DNA of cells in both animals and plants, but because radiation is a constant presence in the environment living organisms have evolved cellular repair mechanisms that constantly act to rebuild DNA and reverse any damage. Even so, one would expect rates of cancer and other ill-effects from radiation to vary by region between places where the received doses are highest. The science says otherwise, however. One study of Ramsar, Iran, where radium in hot springs delivers a hefty dose of natural radiation to the local inhabitants – a hundred times that permitted for radiation workers in nuclear power stations – found no difference between them and people living in more normal areas.[38] A similar

15-year study of residents living in a high-background-radiation area in Yangjiang, China, also found no increase in cancer risk.[39] On the other hand, there does seem to be a small statistical link between rates of naturally occurring radon and lung cancer, though the risk is many times higher for smokers.[40]

Compared with these natural doses, the radiation releases from nuclear power stations are typically very low. A single 10-hour flight would in most cases yield a higher exposure to the average citizen than a whole year's worth of the increased radiation now in the environment due to civil nuclear power. This makes claims by anti-nuclear activists that nuclear power stations are associated with 'cancer clusters' unlikely to be true, if only because the radiation increases purported to cause the extra cancers are so small. This issue has been studied exhaustively by scientists, and the vast majority of studies have found no link between nuclear power stations and cancer incidence in the local populations of nearly a dozen countries from France to Sweden.[41]

The occasional study does seem to find a link, however, and these – rather like the odd piece of science that seems to confound global warming theory – are pounced upon by campaigners. One highly publicised recent study first came to my attention when it was used by the British Liberal Democrat party as support for its anti-nuclear stance before the 2010 election. This piece of work, performed by German researchers and published in 2008, concluded that there had been a doubling of the incidence of childhood leukaemias (though no other cancers) in young people living within 5 kilometres of a nuclear power plant.[42] Specifically, over a 23-year period there were 37 cases of leukaemia when statistically 17 would have been expected – not a very large absolute increase, totalling only one extra case in the entire country per year, but still worth investigating.

With an additional radiation exposure to residents living close to German nuclear power stations that is 10,000 to 100,000 times less than natural background levels, however, the nuclear plants can be safely ruled out as a causal factor in these leukaemia cases – as the authors of the study acknowledge. Given that no additional incidences of any other types of cancer were found, the results were probably nothing more than a statistical accident. This is particularly the case given that natural background radiation levels in Germany vary by a factor of ten around the country, and yet there is no corresponding pattern in leukaemia incidence. It is intriguing that cancer clusters have also been identified in areas where nuclear power stations are planned but never built, where they are built but before they are switched on, and around other large infrastructure projects that do not emit radiation at all.[43] In other words, cancer clusters may be associated in some circumstances with nuclear power stations – but they are almost certainly nothing to do with radiation emissions from them.[44]

The mandate of this book, moreover, is to look at environmental hazards to other species than just humans, with a view to building up a picture of impacts that matter at an Earth-system scale. A new scientific discipline, radioecology, has sprung up to study the effects of radiation in the environment, and its conclusions, on the whole, are positive. Thousands of scientific papers have now been published examining the occurrence and impacts of radioactive isotopes – some natural, some artificial – on biota in a wide variety of different environments, from forests to sand dunes, but none that I have come across have found convincing evidence of harm. My tentative conclusion is that it does not seem at present that artificial radiation from either nuclear power, medical uses or weapons tests is negatively affecting biodiversity anywhere in the world. I cannot think of any way that nuclear power significantly affects any of the other planetary boundaries other than in the area of toxic pollution, so this is indeed reassuring.

Before Fukushima, the one exception to this benign picture was of course Chernobyl. At 1.24 am on 26 April 1986 Chernobyl's Unit 4 reactor exploded after staff disabled safety systems and performed an ill-advised experiment to check – ironically enough – the reactor's safety. The top was blown off the reactor building, and hot pieces of nuclear fuel and reactor core were scattered all over the site, including on the roofs of neighbouring buildings. During the next ten days, dust and smoke from the site released a radioactive plume that spread over most of Europe, seriously contaminating large areas of what is now the Russian Federation, Belarus and Ukraine. Hundreds of thousands of people were evacuated within hours from the surrounding areas, leaving whole villages and towns ghostlike as their inhabitants were moved elsewhere. To date radiation from Chernobyl is still detectable all over Europe from the Swedish Arctic to the English Lake District, and the Chernobyl Exclusion Zone around the damaged reactor is almost entirely uninhabited.

Despite all this, Chernobyl was a long way from being, as some have claimed, 'the world's worst industrial disaster'. Nor was the death toll counted in the tens, or even hundreds, of thousands, as anti-nuclear campaigners and the media have often asserted.[45] Exhaustive studies of affected populations, firemen who attended the blaze (many of whom received colossal radiation doses), and the thousands of 'liquidators' who later cleaned up the site, yield an estimated death toll that currently stands at less than 50. Several thousand children did suffer from thyroid cancer as a result of radioactive iodine doses received from Chernobyl – but as thyroid cancer is relatively treatable, by 2002 thankfully only 15 of the estimated 4,000 cases of childhood thyroid cancer had proved fatal.[46] All this could have been avoided had the Soviet authorities distributed iodine pills to the affected populations.

In 2000 the UN Scientific Committee on the Effects of Atomic Radiation (UNSCEAR) concluded that 'there is no evidence of a major public health impact related to ionising radiation 14 years after the Chernobyl accident. No increases in overall cancer incidence or mortality that could be associated with radiation exposure have been observed.' Moreover: 'The risk of leukaemia, one of the most sensitive indicators of radiation exposure, has not been found to be elevated even in the accident recovery operation workers or in children. There is no scientific proof of an increase in other non-malignant disorders related to ionizing radiation.' Nor is there any scientific medical basis to reports of increased abnormalities amongst children.[47] The well-intentioned activities of the various 'Children of Chernobyl' cancer charities notwithstanding, there is no evidence for claimed increases in deformities or illnesses in children exposed as compared with unexposed populations in eastern Europe.

None of this is to downplay the terrible impacts of the Chernobyl disaster on the people who were affected by it. But the most serious impacts of Chernobyl may actually be social and psychological: higher levels of depression and anxiety have been reported in people who were exposed to radiation or evacuated – and this stress seems to have been passed from parents to children. People's self-identified status as 'Chernobyl victims' has led to dependency, poor health, alcohol abuse and even suicide, according to UNSCEAR.[48] The unfortunate truth is that the general post-Chernobyl anti-nuclear hysteria, reinforced by exaggerated death tolls and impacts published over subsequent years by environmental groups, has probably worsened the victim status trauma suffered by the people who lived in the area. Indeed there is strong evidence that fear of radiation has been much more damaging than radiation itself: consider, for example, the large number of additional abortions undertaken by women in Eastern Europe who considered themselves to have been exposed and therefore likely to bear deformed or diseased children.

Yet the emotive impact of anecdotal evidence from Chernobyl has been used relentlessly by environmental groups in support of their battle against nuclear power. Greenpeace maintains a website of heart-rending black and white photos of mentally retarded, cancer-stricken and deformed children, taken in the Ukraine and nearby countries, with the title 'The real face of the nuclear industry'.[49] Given the lack of these claimed horrors found by scientifically authoritative studies of Chernobyl, this kind of propaganda seems to me to be an abuse of real people's suffering. For example, I find the statement of Greenpeace International's director Gerd Leipold, made on the twentieth anniversary of the Chernobyl disaster in 2006, quite distasteful. Leipold laments 'the bedridden children with cancers and degenerative diseases who must be turned every fifteen minutes in excruciating pain', and 'the parents who themselves suffer from chronic radiation-related diseases' and concludes on this basis that 'nuclear power is inherently highly dangerous' and that 'another catastrophe on the scale of Chernobyl could still happen anytime, anywhere'.

This assertion ignores the fact that the Chernobyl-type reactor, aimed primarily at producing plutonium during the Cold War, was inherently dangerous and that none were built outside the Communist bloc. Using similar tactics to climate-change deniers, Greenpeace in 2006 handpicked a group of supportive scientists in order to publish its own 'scientific' report challenging the expert consensus around Chernobyl, insisting that 100,000 fatal cancers could be expected and that the mainstream scientific effort was trying to sweep evidence under the carpet.[50]

In his twentieth-anniversary statement, Greenpeace's Gerd Leipold invited his audience to remember 'the old people who have no alternative but to eat mushrooms and burn firewood harvested from woodland so radioactive that soil samples from them are treated as radioactive waste in Western Europe'. Lest we forget, 'it is here where we should look – into the eyes of these people – when we are told about the so-called "benefits" of nuclear power.' As it happens, on a visit to Chernobyl in May 2010 with a Channel 4 camera crew, I had the opportunity to look into the eyes of just such an old person during an interview I conducted with 81-year-old Leonid Petrovich outside the garishly painted church in Chernobyl village. Despite having moved back to his old cottage deep inside the exclusion zone, Petrovich did not at all resemble the sad victims portrayed by Greenpeace – and nor did any of the other people, mostly old and all Chernobyl residents, whom I also spoke to on the same day and who were accompanying him to church. All had moved back of their own accord, and none were particularly concerned about the dangers of radiation, which they considered to have been over-hyped by the authorities. As a milestone in his life, Leonid Petrovich told me that the explosion of Reactor Number 4 in April 1986 was a great deal less traumatic than the arrival of German SS troops in 1942, when Petrovich was a boy, and saw dozens of Jewish residents shot in the forest outside the village.

Poking around the Chernobyl exclusion zone was a fascinating experience for me. In case readers are concerned, the radiation dose I received over the two days I spent there was about 0.1 milliSieverts, 100 times less than a single medical CT scan.[51] What most struck me was the sheer profusion of wildlife. The long sleeves, strange-smelling wristbands and hats we had been ordered to wear turned out to have nothing to do with radiation – they were to ward off the clouds of mosquitoes that emerged from every bush and tree in the lush new forest. Just a couple of miles from the reactor itself, in the abandoned Soviet town of Prypiat, I variously saw a snake, a lizard, and a large hare lolloping past the rusting dodgems of an abandoned funfair. The birdsong was nearly deafening. Everywhere wildlife was verdant and abundant – hardly the image I had grown up with of a devastated, blighted landscape. A silver birch tree – always a hardy pioneer species – was growing up through the wooden floorboards of an abandoned gymnasium. The calls of cuckoos echoed between the abandoned apartment blocks.

None of this is to suggest that radioactivity is somehow good for biodiversity. What has benefited the local wildlife is one thing only: the exclusion of people. Moreover, the Chernobyl disaster did at first have a serious effect on the surrounding ecology. Nearby pine trees quickly died, becoming the famous 'red forests' as their needles shrivelled and turned brown. Within a radius of about ten kilometres, other trees developed malformed buds, dead branches or genetic abnormalities. Researchers also noticed a 'catastrophic impact' on soil invertebrates: most of the worms and insects living on and below the forest floor quickly died off as the radionuclides were washed into the leaf litter by rain. Scientists found that rodent populations had suffered, their numbers declining and reproductive success compromised. Fish in heavily irradiated water bodies like the nearby Chernobyl cooling pond were also affected, suffering problems with spawning and fertility – those fish that received the highest doses were even made sterile. Domestic animals kept on nearby farms – including cattle, chickens and sheep – also showed the impacts of severe radiation damage.[52]

But what is most striking about the ecological impacts of the Chernobyl disaster is how limited they were. There is little evidence of negative impacts on biodiversity anywhere except within a very small area immediately surrounding the reactor that received the highest doses of radiation. Some classes of animals also seemed relatively unaffected: wild birds were raising apparently healthy chicks in the area within months. Moreover, the general environmental recovery was spectacular: forests began to regrow within a year, whilst scientists studying the soil invertebrate fauna found a 'total recovery' of biomass within two and a half years, and in species diversity within a decade. Ecological reports compared the effects of Chernobyl on the surrounding ecosystems as being similar to those of a forest fire: dramatic collapse and death, followed by rapid renewal. Today the eerie 'red forests' are gone, replaced by vigorous new growth of pine and birch.

Scientists working in nearby woodlands in 1994, less than a decade after the disaster, were amazed at how unaffected wild animals seemed to be. 'During our excursion through the woods, we trapped some of the local mice for examination in a makeshift laboratory,' write US biology professors Ronald Chesser and Robert Baker. 'We were surprised to find that although each mouse registered unprecedented levels of radiation in its bones and muscles, all the animals seemed physically normal, and many of the females were carrying normal-looking embryos. This was true for pretty much every creature we examined – highly radioactive, but physically normal.'[53] Chesser and Baker found they had to ditch some cherished assumptions, even being forced – in great embarrassment, after several re-evaluations of the original data failed to replicate the results – to retract a 1996 paper they had published in *Nature* claiming that voles living close to Chernobyl had an elevated rate of genetic mutation. Actually, the voles had shown no impact at all. 'It was an important lesson in admitting error and coming to terms with our mistakes,' the chastened duo admitted later.

Those people who have returned to live in the Chernobyl Exclusion Zone frequently tell outsiders how they now have to put up with an annoying profusion of wildlife. Leonid Petrovych reeled off the list to me as we sat outside Chernobyl village church: deer, wolves, wild boar, eagles, all confirmed in biodiversity studies by scientists. In a recent harsh winter the resurgent wolf packs had even killed and eaten some of the local domestic dogs, and attacked a local man as he sat playing cards. The exclusion zone has now become a favoured breeding area for white-tailed eagles, spotted eagles, eagle owls, cranes and black storks, and in total 50 endangered species of birds, mammals, reptiles and amphibians now inhabit the abandoned town of Prypiat and its environs. Recognising the growing wildlife value of the zone, managers introduced 28 endangered Przhevalsky wild horses in 1998, and within six years their numbers had doubled. The conclusion is inescapable: even the worst-case nuclear accident, scattering intense radiation over a wide area, is better for biodiversity in general than normal, everyday human habitation. 'Without a permanent residence of humans for 20 years, the ecosystems around the Chernobyl site are now flourishing,' a 2006 scientific report published by the International Atomic Energy Agency concludes.[54]

Again, I am not suggesting that biodiversity has benefited from radiation, except as a result of its unintended consequence of reducing everyday human pressures on wildlife. Indeed, some scientists who have studied animal populations in the exclusion zone suggest that biodiversity is lower in areas where contamination levels are highest. Two ecologists, Tim Mousseau and Anders Möller, have published several papers reporting declines in biodiversity – from insects to birds to mammals – which they claim correlate with areas of higher radioactivity.[55] However, these results are highly controversial in the scientific community,[56] and Moller in particular has been subject to accusations of bias and even fraud.[57] In 2001 the Danish Committee on Scientific Dishonesty found against him in a separate case regarding a paper he had published on oak leaves. A 2011 study by the same authors that claimed to correlate smaller bird brain sizes with higher radioactivity also seems to me to be suspect:[58] the supposedly observed effect is tiny, just a few percentage points over a 10,000-fold difference in radioactivity, and likely a statistical artefact. If, as I think is likely, the effects of radiation on wildlife do turn out to be minimal, there are obviously better ways to create nature parks than conducting ill-advised experiments on badly designed nuclear reactors. But the experience of Chernobyl supports my assertion throughout this book that civil nuclear power is not a serious threat to ecosystems, even when something goes badly wrong – as it so obviously did on 26 April 1986.

Something also went badly wrong in Japan on 11 March 2011: in this case a magnitude 9.0 offshore earthquake followed by a 14-metre tsunami, which swept into coastal towns as a catastrophic black surge, tearing apart buildings and scattering cars in scenes of unparalleled devastation. The wave also struck the Fukushima Daiichi nuclear plant, which – although its operating reactors had shut down safely during the earthquake itself – required electrical power to maintain cooling of the reactor cores and nearby ponds storing spent nuclear fuel rods. Backup diesel generators were swamped by the wave, and without electrical power three of the reactors began to go into meltdown. This is about the worst thing that can happen to a nuclear plant, and the Fukushima workers risked their lives heroically to maintain some level of cooling – pumping in sea water, bringing in fire trucks, and at one point even calling in the army to drop water from helicopters in a desperate bid to maintain water levels in the spent fuel ponds. But at the time of writing (early April 2011) it looks as though triple partial meltdowns were suffered, one of the containment vessels may even have been breached, and the entire plant will have to be decommissioned at enormous expense.

It is also undeniable that significant releases of radioactive materials took place, in particular in the immediate hours and days after the disaster when steam building up in the overheating reactors had to be vented in order to prevent pressure in the reactor cores building up to dangerous levels. This steam contained hydrogen, which exploded and largely destroyed three of the reactor buildings, further worsening the situation. Spikes of radioactive iodine were measured as far away as Tokyo, and in the weeks that followed, significant contamination also took place in coastal waters as radioactive seawater flowed back out from the stricken plant. Even plutonium particles – probably from one of the spent fuel ponds – were identified nearby. It was also reported that two workers wading through radioactive water in one of the reactor buildings suffered significant and potentially dangerous radiation doses to their legs. This was not Chernobyl, but it was far and away the worst disaster to have hit a civil nuclear plant outside the former Soviet Union, worse even than the Three Mile Island partial meltdown in 1979 which so spooked America.

But context is all, and within the context of a tsunami disaster that likely killed 28,000 people, Fukushima's death toll is still – and will likely remain – zero. The increased levels of radioactive iodine measured in Tokyo tapwater sounded scary, but even at the height of the crisis were far below legal limits in Europe. Similarly, contaminated seawater offshore from the plant was presented in the media as an ecological nightmare, but there is highly unlikely to ever be any identifiable impact on sea life. Nor should humans be affected. As the Fukushima operator pointed out, a member of the public would have to eat seaweed and seafood harvested just one mile from the discharge pipe for a year to receive an effective dose of 0.6 millisieverts: less than a quarter of the 2.7 millisieverts each of us absorbs each year from natural background radiation, with no apparent ill effects. Iodine 131 also has a half-life of only 8 days, so quickly disappears from the scene. Within the confines of the Fukushima plant there are other isotopes which present much greater long-term challenges, but contamination over wider areas is sufficiently low to enable people in the evacuated zone to return and continue their lives without fear of long-term health problems due to radiation. Indeed, I would expect the toxic contamination from the tsunami itself – which swept through industrial infrastructure, fuel tanks and so on – to present a much more serious problem in the affected areas.

I repeat: context is everything, and no energy source is without danger. People die in industrial accidents making steel for wind turbines, and fall from roofs whilst installing solar panels. Some analysts have conducted assessments of 'deaths per terawatt-hour' which find nuclear to be even safer than renewables as an energy source.[59] Even energy efficiency has its downside: building accidents mean that energy savings from draught-proofing Swedish buildings apparently cause more deaths than a similar amount of energy produced by nuclear or hydropower.[60] I have limited confidence in these comparative exercises, but it is surely beyond doubt that fossil fuels – of every type – are far more dangerous than nuclear. As a writer at the *Atlantic* magazine concluded: 'We need to take nuclear safety concerns very seriously, but let's not forget what the baseline for our energy system is.'[61]

The baseline for 2010, the article pointed out, contained the following energy-related disasters: 7 February – a refinery explosion in

Connecticut killed 6; 15 March – a coal mine fire in Zhengzhou, China, killed 25; 20 March – another coal mine collapse in Quetta, Pakistan, killed 45 miners; 28 March – a coal mine flood killed 38 in Shanxi, China; 31 March – 44 workers died in a mine explosion, in Yichuan, China; 2 April – an oil refinery blast in Washington State led to 5 deaths; 5 April – 29 miners were killed in an explosion in a West Virginia mine ... the published list includes 25 serious accidents causing multiple fatalities for 2010 alone, none of which made the news where I live. Presented in this context, Fukushima was a moderate industrial accident – nothing more. It is certainly no argument against the continued, and increased, use of nuclear power, particularly as newer reactor designs mean a much greater degree of passive safety can be employed in future than was the case for the 1960s-era boiling water reactor designs deployed at Fukushima Daiichi.

Fukushima and nuclear safety aside, what about the problem of nuclear waste? To quote Friends of the Earth this time: 'Despite more than half a century of nuclear generation, it is still not known how to safely manage its toxic legacy.'[62] That radioactive waste is an 'unsolved problem', posing environmental risks far into the future that we don't know how to deal with, is a central argument of the anti-nuclear lobby. Yet nuclear waste is actually only an 'unsolved problem' because anti-nuclear campaigners do not agree with the proposed solutions. These solutions are quite simple: once spent fuel rods are removed from the reactor core, they are stored in cooling ponds until their radiation levels decline sufficiently for them to be stored in dry steel casks. The levels of radioactivity emitted decline by a thousand times in 40–50 years. In the longer term, geological disposal of waste that cannot be recycled or otherwise put to good use (which the vast majority can) is a straightforward engineering challenge that poses negligible risks over the long term. Few people seem to realise that the radioactivity of nuclear waste declines with time; and the more radioactive the waste is to start with the more quickly the levels of radiation decline.

Moreover, the dangers and timescales involved are routinely exaggerated. Friends of the Earth offers as a reason for its opposition the claim that nuclear waste remains 'dangerous for tens of thousands of years'.[63] This is technically correct, but misleading nonetheless. The vast majority of waste will be no more radioactive than the natural uranium ore that it was originally derived from in just a few hundred years.[64] And the isotopes that stay radioactive for longer (like isotopes of plutonium) can be recycled in other reactors, so need not be buried at all. In comparison, of course, hazardous wastes containing toxins like mercury and arsenic stay dangerous for ever – and yet are barely controlled in much of the world, and attract nothing like the passion that the nuclear issue does. The volumes of nuclear waste are also tiny compared with other competing technologies. A 1000-megawatt nuclear reactor produces less than 30 tonnes of used fuel per year, most of which can be reprocessed and used again. In comparison, a coal-fired plant of the same capacity will discharge 400,000 tonnes of ash and several million tonnes of carbon dioxide over the same year.[65] Even after a quarter-century of using nuclear power to produce 80 per cent of its electricity, France is still able to keep all the high-level waste generated by all its nuclear plants under the floor in a single room. This is not an 'unsolved problem'. It is not really much of a problem at all.

All of this would be academic were nuclear not an essential part of any realistic plan to meet the climate change planetary boundary. As this book has shown, renewables are a crucial part of our toolkit, but not enough on their own. The battle of the energy titans comes down to one great contest: nuclear vs. coal. And by rejecting nuclear over past decades Greens have unwittingly kept the door open for this most polluting energy source of all. For example, several planned reactors in the United States, after being stridently opposed by Greens in the 1970s and 1980s, became coal stations instead. In Austria, after anti-nuclear activists won a nationwide referendum in 1978, a whole country turned from nuclear to coal: and an entire completed nuclear power station was pointlessly mothballed right after being built.

An interesting 'what-if?' exercise arises. What might be the quantity of carbon dioxide emitted over the last few decades from fossil-fuelled power plants as an accidental by-product of anti-nuclear campaigning? In Austria, for example, six nuclear stations were proposed, and none were eventually used. In the US, at least 19 nuclear plants were cancelled after being proposed – mainly due to the changing tide of public opinion brought on by the rise of the Greens. What if the nuclear build rate of the 1960s and 1970s had continued until today, and all these proposed plants had been welcomed by the rising environmental movement? There can of course be no definitive answer to such a question, but if we say that 150 additional plants would by now have been running for 20 years, these would have avoided the emission of 18 billion tonnes of CO_2.[66] In climate-change terms, opposing nuclear was a gargantuan error for the Greens, and one that will echo down the ages as our globe's temperature rises. Some in the environmental movement have begun to realise this mistake, including members of the Green Party and the former director of Greenpeace UK, Stephen Tindale, who courageously joined with me to make a front-page 'mea culpa' declaration in the *Independent* newspaper on 23 February 2009.[67] In the US, both Stewart Brand and NASA scientist (and planetary boundaries co-author) James Hansen have strongly supported nuclear in the battle against climate change. In Britain, my friend and colleague the writer George Monbiot, one of the Green movement's most fearsome and well-known campaigners, wrote in the *Guardian* that the Fukushima disaster had convinced him that nuclear power was actually *less* dangerous than his environmental comrades confidently asserted, especially when compared to fossil fuels. For Monbiot the global warming argument was crucial. He wrote: 'Like most environmentalists, I want renewables to replace fossil fuel, but I realise we make the task even harder if they are also to replace nuclear power.'[68]

I should also emphasise that the Green movement has in recent years put vastly more effort into opposing coal than nuclear, with large numbers of coal plants in the US and Europe now cancelled as a result of the efforts of modern environmentalists. According to the Sierra Club, by February 2011 150 proposed coal plants had been cancelled in America alone since 2001, with only 41 left in development or construction.[69] But there is no point in opposing coal, as the Sierra Club and others have done so successfully, if you also oppose its main alternative, and regarding nuclear I have yet to see much sign of a shift within the more established Green NGOs, from Greenpeace and Friends of the Earth in the UK to the Sierra Club and NRDC in the US, still less within the official Green parties of Europe and elsewhere. In a particularly dispiriting example of this ideological stasis, the Green Party in Germany won an unprecedented victory in a regional election on 27 March 2011 after campaigning partly on the basis of public fears about Fukushima and nuclear power.

What would be most tragic, however, would be for a new generation of young Greens to be indoctrinated with the unscientific anti-nuclear prejudices of the old. In January 2011 Greenpeace produced a glossy report aiming for the total elimination of nuclear in Europe by 2050 – and its replacement in the short and medium term with the fossil fuel gas.[70] I hope this book makes a clear case that environmentalism must change, and on the nuclear issue more change is needed than perhaps in any other area. Leaving behind strong ideological commitments is always painful, but in a changing world is often necessary. If we are to properly address the challenges posed by the planetary boundaries, environmentalism will have to raise its game.

CHAPTER EIGHT

The Aerosols Boundary

We live on a smoky planet. Viewed from space, the Earth is brighter than it once was, as sunlight reflects outwards from a worldwide haze layer several kilometres thick. From the ground, this haze makes skies that were once an intense dark blue appear milky over most of the planet's inhabited areas.[1] The denser the population, the thicker the haze layer: one pollution plume streams out into the Atlantic from the east coast of North America, whilst an equivalent European plume spreads eastwards into Asia. From China a thicker pall stretches across the Pacific, whilst smoky clouds from South Asia and southern Africa extend over the northern and southern portions of the Indian Ocean. Worldwide, smoke, dust and other airborne particles released by human activity have the same combined effect as a constant medium-sized volcanic eruption, scattering sulphur and soot high into the upper atmosphere. Here is another, very visible, manifestation of the Anthropocene: we have changed the colour of the sky.

Together, these anthropogenic airborne particulates are known as 'aerosols', and their impacts jointly form one of climate science's greatest unsolved puzzles. The same particle may have a warming or a cooling effect overall, depending on its elemental makeup and where it hangs in the atmosphere at any precise time. In general sulphate particles are bright in colour, and therefore reflect sunlight, whilst soot is dark and absorbs it. But aerosols can also act as nuclei for cloud droplets to condense around, forming white clouds that – like snow and ice – reflect incoming solar radiation. Darker dust and soot, however, make duller clouds, which can absorb more heat and therefore be faster evaporated than their whiter cousins. The precise effect of aerosol haze, in addition, depends on what is underneath it. If it screens sunlight hitting dark-coloured ocean, it will have a cooling effect. If it dims sunshine heading towards the highly reflective ice caps, on the other hand, it could warm the planet overall. Confused? So are the scientists. And so, consequently, are the climate models they build, which can give very differing interpretations of the combined impacts of anthropogenic pollution on climate. The current scientific consensus is that, taken together, aerosols have a substantial overall cooling effect. But its exact magnitude is uncertain.

What is certain, however, is that aerosols all have a very short lifetime. Unlike carbon dioxide, which stays in the atmosphere for decades or even centuries, dust, smoke and sulphur particles are washed away by rain in a matter of days. So whilst the greenhouse gas burden now changing the climate results from accumulated carbon emissions over more than two hundred years, aerosols circulating in the air were released at most only a week or two ago. This is a problem that only persists as long as we go on causing it. Consequently, it will be one of the easiest planetary boundaries to address. Although the planetary boundaries expert group – as a result of the tremendous uncertainties attached to aerosol pollution's impacts on the climate and Earth system in general – has been unable to come up with a specific quantified limit to atmospheric concentrations of any specific pollutants, this hardly matters. It is clear that having clean air will produce a lot more benefits than drawbacks. Moreover, we can clean up our activities relatively quickly and cheaply, with readily available technology. Additionally, this is a problem that will solve itself as developing countries become more prosperous and their populations demand pollution reductions. The challenge is to solve the problem in a way that complements, rather than conflicts with, the other more challenging planetary boundaries.

FIRE, FLOOD AND ICE

Perhaps the most famous pall of pollution anywhere in the world sits over India. The so-called Asian Brown Cloud is the combined product of thousands of coal-burning power stations, tens of thousands of factories and millions of open fires burning biomass like wood and dung, and is now a semi-permanent feature of the South Asian meteorological map. By reducing solar radiation at the surface, it has cooled the entire north Indian Ocean and changed the dynamics of the monsoon, perhaps the world's most spectacular and important meteorological phenomenon. At its height in the spring, the brown cloud creeps north up to the ramparts of the high Nepali Himalayas, its sooty deposits darkening the snow on the flanks of Mount Everest itself.

If China is included too, rapidly industrialising Asia is the densest source of global pollutants. Scientists at high-altitude monitoring sites on Mauna Loa volcano in Hawaii, and at Boulder, Colorado, in the Rockies, puzzled at first by the rising levels of sulphur their laser measurements found in the stratosphere, have discovered that Chinese coal-fired power stations are now a leading source of sulphur reaching the high stratosphere.[2] If Chinese sulphur emissions double over the next couple of decades as predicted, they will be equivalent to 5 per cent of the 1991 Pinatubo eruption – the second-largest volcanic explosion of the entire twentieth century. This will, experts calculate, reduce temperatures in the lower atmosphere by 0.03°C and will even affect the ozone layer – truly a worldwide impact. Aerosol pollutants from India are hoisted 15 vertical kilometres by powerful updraughts during the country's violent monsoon rainstorms, allowing them too to circumnavigate the globe.[3]

Those who suffer most from the direct impacts of this pollution, of course, are those living in Asia. Airborne particulates are a leading cause of heart and lung disease throughout the continent, where 3 billion people breathe air classed as dangerous by the World Health Organisation.[4] The Chinese Academy for Environmental Planning blames air pollution for 411,000 premature deaths each year across the country, probably a serious underestimate given the year-round smog pall hanging over most Chinese cities. In Beijing, where 1,200 new cars and trucks are added to the capital's roads each day, deadly particulates known as PM 10s average around three times the WHO standard.[5] Rich countries are affected too, though to a much lesser extent: in the UK the Parliamentary Environmental Audit Committee recently concluded that up to 50,000 people could be dying prematurely each year due to air pollution,[6] whilst for the US, the American Lung Association's *State of the Air 2010* report states that 'over 175 million people – roughly 58 per cent – still suffer pollution levels that are too often dangerous to breathe'.[7] Children are particularly vulnerable to traffic-exhaust emissions, with impacts ranging from respiratory ailments like asthma and chronic bronchitis to abnormal lung development and even cancer. Yet studies show that a third of US schools are located within 400 metres of major roads, well within the traffic pollution danger zone.[8]

Because of their global distribution and effects on solar-heat absorption and distribution, atmospheric aerosols are having major impacts on the Earth's hydrological cycle. Pollution from Northern Hemisphere smokestacks is thought to have played a significant role in triggering the terrible decades-long drought that struck the Sahel region of Africa in the 1970s and 1980s, which led to regional famines with death tolls in the millions.[9] As sulphur and soot darkened the atmosphere above the North Atlantic Ocean, they pulled life-giving rains away from North Africa, allowing the desert to spread south as precipitation levels declined by 40 per cent.[10] Thanks to pollution controls, the trend has now reversed, and rainfall over the Sahel largely recovered.[11] The reduction in 'global dimming' that helped prevent the Sahel turning into permanent desert has in fact increased rainfall right across the globe, with an upward trend of more than 30 mm measured between 1986 and 2000.[12]

Using sophisticated models run on powerful supercomputers, scientists have begun to work out the ways aerosol emissions are affecting the Indian monsoon. First, it seems, a 10 per cent drop in sunshine reaching the ocean lowers the level of evaporation, reducing the moisture available for rainfall. Second, the Asian brown cloud – thickest over the densely populated landmass of India – also reduces the temperature difference between land and sea that is the main engine of the monsoon. The differing impacts on the northern and southern Indian Oceans – where the northern half cools, and the southern half, which remains under relatively clean air, warms – also hampers the monsoon circulation.[13] Over much of India, Bangladesh,

Burma and Thailand summer monsoon rainfall has been declining, as brown-cloud aerosols weaken circulation patterns that sustain the food production and livelihoods of over a billion people across the subcontinent.[14] In China, aerosols have split the country in half, bringing drought to the north and floods to the south by displacing the usual annual progression of the East Asian monsoon.[15]

Like toxins, atmospheric pollutants tend to concentrate in the frigid, stable atmosphere of the Arctic, where as long ago as the 1950s air pilots noticed a whitish haze obscuring the horizon. So-called 'Arctic haze' has since become recognised as a well-established phenomenon, one that may be making a significant contribution to atmospheric warming.[16] Likely sources are industry and coal-burning in Eurasia, plus occasional fierce wildfires like those that swept Russia in the summer of 2010.[17] As the soot from the haze layer gets gradually deposited on the snow, it darkens the surface and adds to springtime melt. One study suggests that sooty deposits in Northern Hemisphere snow add a fifth to the melt-rate, and are a major contributor to Arctic warming.[18] The effect has been observed too on the other side of the world, where smoke from the burning Amazon rainforests is darkening the snow over faraway Antarctica.[19]

In the Himalayas, jammed between the world's two largest sources of soot – India and China – the glaciers that cloak the world's highest mountains are getting increasingly dirty. Scientists working at the world's highest atmospheric research station, 5,000 metres up near Everest Base Camp above the Khumbu glacier, have measured increasing levels of pollution streaming up the valley from densely populated regions far below,[20] forming brown clouds sometimes thick enough to cause several degrees of local warming.[21] This darker snow absorbs more heat from the sun – as much as double in some conditions, accelerating the melt of the glaciers and starting the melt season earlier in the year.[22] The predicament of the Himalayan glaciers is not unique to Asia: in the European Alps, the continent's diesel engines are thought to be depositing enough soot on glaciers to double their absorption of solar heat and rapidly accelerate the ongoing melt driven by global warming.

BLACK CARBON

Whilst overall aerosols almost certainly have a profound cooling effect on the climate, one type of aerosol particle works strongly the other way. In fact, it may be the second-largest contributor to global warming after CO_2. This is the humble soot particle, known by atmospheric scientists as 'black carbon'. Per unit of mass, black carbon is a million times more absorbent of heat than carbon dioxide,[23] and even though it lasts in the atmosphere for only a few days, averaged out over a century it is – pound for pound – five hundred times more effective in warming the globe than CO_2. By cleaning it up, we can not only improve respiratory health for human populations around the world, but we can also avoid a quarter to a third of ongoing global warming.[24]

Reducing black carbon emissions could immediately help stabilise both the polar ice caps and rapidly melting mountain glaciers. This is because, as I mentioned above, soot deposited on snow and ice makes the surface darker, accelerating melt rates. The importance of this can scarcely be overstated. As Stanford University's Mark Jacobson puts it: 'Controlling soot may be the only method of significantly slowing Arctic warming within the next two decades,' and thereby avoiding a runaway collapse of the Arctic sea-ice sheet.[25] Because of the different timescales involved, dealing with black carbon is not a substitute for reducing greenhouse gas emissions, but it is a crucial accompaniment. Unfortunately, the issue has so far received very little attention either from campaigners, the media or policymakers. This has to change, and quickly, because the world is missing an easy opportunity for a win–win environmental outcome that could relatively cheaply reduce the near-term warming of the globe at the same time as avoiding an estimated 1.5 million annual deaths per year from air pollution.

The challenge of black carbon should be an easy one to address, because nobody actually wants it. Unlike carbon dioxide, which is an unavoidable consequence of any use of fossil fuels for energy production, black carbon is only generated by incomplete combustion. We have all seen it, in the clouds of black smoke emitted by badly maintained diesel engines, or the sulphurous fumes of a home coal fire.

The emission of black carbon is therefore easily avoidable by the improvement of combustion, the capture of particles, or the substitution of fuels. Importantly, black carbon also challenges the conventional narrative of global warming that sees rich countries as culprits and poor countries as innocent victims. Because it is typically produced by low-tech combustion, two thirds of the world's soot comes from developing nations, making them nearly as responsible for day-to-day climate change as the industrialised world.[26] Indeed, as much as a quarter of the global annual output of black carbon comes from just two countries: India and China.[27]

Some major sources of black carbon, like wildfires and other uncontrollable burning in open areas, we can do little about. But its production could be reduced by up to three-quarters by taking some relatively simple measures. In rich countries, soot comes mainly from diesel engines: this can be avoided by the mandatory sale of cleaner diesel fuel and the fitting of 'diesel particle filters' on the engines of buses, trucks and diesel-burning cars. Like the installation of catalytic converters on petrol cars, which only became widespread when made mandatory by legislation, the time has come to force diesel vehicles too to clean up their act. In the European Union, legislation is already pushing rapidly in this direction: since January 2011, all new cars must comply with much stricter emissions standards (known as Euro V), which will almost certainly make the fitting of diesel soot traps standard. But retrofit regulations are also likely needed, especially for large vehicles, to make older smoky trucks and buses as clean as newer models. London's 'Low Emission Zone' is aimed at encouraging these retrofits, by charging lorries a hefty fine – currently set at up to £200 – if they exceed emissions limits.

The filthiest fuel of all, and a major global source of black carbon, is burned offshore, by the tens of thousands of merchant ships that ply our seas. Emissions in port areas can be significant, and one estimate of global mortality from these deadly particles and other ship-emitted pollutants totals 60,000 people per year[28] – more than a dozen annual Chernobyls.[29] Because their emissions largely take place out of sight, shipowners save money by burning heavy oil in the engines of their vessels, a fuel that is literally the scrapings of the oil-refining barrel. Once again, changes are afoot: the International Maritime Organisation has set a 2020 deadline for big reductions in sulphur and soot pollution from the world's 50,000-strong shipping fleet.[30] Much of this will need to be supplied by cleaner fuels, but on-board scrubbers are also an option that should be mandated in bigger ships to clean up exhaust emissions before they are released. Over the longer term, as fossil fuels are gradually phased out in all forms of surface transport to address global warming, co-benefits in pollution reduction will be realised. I outlined in an earlier chapter that electrification is probably the best option for the world's vehicle fleet: electric cars of course have zero emissions of soot as well as CO_2.

China is not as far behind in its transport pollution-control standards as many Westerners, familiar with horror stories about the country's smoggy cities, assume. The Chinese government bases its vehicle emissions rules on European standards, and the 2005 Euro IV standard now applies nationally for new cars. In India, similar rules are mandated, though only in cities. China may move up to the higher Euro V standards as early as 2012, according to news reports. But the biggest source of Chinese black carbon emissions is not cars but coal, particularly coal burned on domestic fires and in small-scale industries in rural areas. For the latter issue, this is a problem that will be increasingly solved by modernisation, as China abandons old artisanal methods in industries like brick kilns, local power stations, coke and cement production and replaces them with larger-scale centralised industries that have better pollution control. The rapid pace of industrialisation is rightly blamed for causing appalling air pollution in China, but as prosperity grows so both the political will and financial resources are rallying to reduce these impacts at source. In recent years Beijing has implemented stricter emissions standards nationwide for coal-burning power plants and cement production, whilst smaller, dirty factories have simply been closed. Even though coal generation doubled between 2001 and 2006, emissions of soot particulates fell by almost a quarter thanks to modern standards being applied at power stations.[31]

In most of developing Asia, the largest-scale black carbon problem should be the cheapest and easiest to solve. This is the production of smoke from dirty cooking stoves in households. Indoor smoke pollution from old-style stoves or open fires burning wood, dung or coal kills 1.6 million people a year due to respiratory infections worsened by smoke inhalation; India alone suffers as many as half a million premature deaths. Levels of toxins in indoor air can be several hundred times World Health Organisation safety standards. Women and children, who spend most time in the house, are the worst affected. Although the health benefits alone more than justify aggressive action on this problem, it has been estimated that black carbon emissions from South Asia could be reduced by as much as 80 per cent if old-style cooking over indoor open fires were replaced with smoke-free cookers.[32] What is needed is decisive intervention to bring clean cookers and improved fuels to more than half a billion households worldwide that urgently need them.

In order to achieve this, of course, money must be raised. Climate financing is the obvious opportunity: with billions of dollars now changing hands in carbon markets, clean cookstoves could be funded for their climate benefits (both in carbon dioxide and other pollutants) whilst likewise promoting the health of millions in poorer countries. One way of raising funds is via carbon offsetting. Instead of spending money *in situ* to reduce emissions, many consumers and industries – especially in rich countries – would find it cheaper and easier to pay for emissions reductions elsewhere. Financially, this is a no-brainer. Reducing global warming by funding clean cookstoves costs as little as a dollar per tonne of carbon equivalent.[33] For comparison, a tonne of CO_2 was trading for most of 2010 at 14 euros per tonne, whilst the cost of avoiding a tonne of carbon dioxide by putting solar panels on rooftops in Germany has been estimated at 700 euros or more.[34] Using carbon-offset funds to pay for clean cookstoves worldwide is extremely good value for money in terms of climate protection, and could potentially save three lives per minute in developing countries due to the accompanying health benefits for women and children.

Here, unfortunately, we run up against what may be the Green movement's second-greatest climate-change mistake, after its opposition to nuclear power. Carbon offsetting was completely derailed as a climate-mitigation strategy in its early stages by vociferous opposition from environmentalists. Prominent Green writers compared offsets to the 'indulgences' sold by the medieval Catholic Church, arguing that the only result would be the salving of rich people's consciences. 'The sale of offset indulgences is a dead-end detour off the path of action required in the face of climate change,' argued a publication by a Green-minded group called *Carbon Trade Watch* in 2007.[35] The leftist magazine *New Internationalist* wrote in an editorial that carbon offsetting was 'a falsehood – a con'.[36] A clever activist website called Cheatneutral.com promoted the analogy of cheating on your partner: the joke was that if you could pay someone else to remain faithful, then the overall amount of fidelity in the world would remain the same. The website claimed that the cheating analogy was apt because 'in the same way, carbon offsetting tries to make it acceptable to carry on emitting excess carbon'. The crux of the argument was a psychological one: 'If the carbon offsetters persuade you that it's possible to offset your emissions, you'll carry on emitting excess carbon through your lifestyle rather than think about reducing your emissions.'

Unfortunately the Greens got their psychology wrong. They were very successful in establishing the idea in most people's minds that carbon offsetting was a con and a waste of money. But they were spectacularly unsuccessful in convincing the same people that therefore they needed to fly less or otherwise reduce their personal carbon emissions. What happened instead was that people carried on flying, but stopped offsetting. The net effect for the atmosphere, therefore, was that more carbon was emitted than would otherwise have been the case, and additional future warming will be caused as a result. The environmental movement fell into the trap of making the perfect the enemy of the good, and the climate – plus the health of millions of women and children in developing nations – lost out. The lesson here is twofold. First, guilt-tripping doesn't work as a campaigning strategy. If you make people feel bad about what they do, you must give them a realistic and feasible alternative. Second, pragmatism beats purism. Every time.

Some caveats are again in order. I have no way of knowing how influential Green critics actually were in holding back the rise of the voluntary carbon market, or how many millions of stoves offsetting companies – like JP Morgan ClimateCare, which runs a successful stoves project in Uganda – would have been able to install if the environmental movement had been supportive from the beginning. My impressions are anecdotal not empirical. Globally, the voluntary carbon market's overall value actually fell between 2008 and 2009, from $419 million to $338 million,[37] but this drop was probably mainly down to the economic recession. Also, some of the early criticisms were correct in that not all carbon offsets sold – particularly those based on questionable tree-planting schemes – were likely to mop up all the carbon emitted elsewhere.

Some environmental groups also deserve praise for playing a constructive role in helping the new industry meet the highest standards: WWF, for example, was involved in designing an industry 'gold standard' for carbon offsets. But I still often receive emails from people who are worried that they might be ripped off and want to ask whether offsetting is worthwhile. My answer is an unhesitating 'Yes': anyone – everyone – who cares about the climate should offset the entirety of their emissions each year (not just those from flying), and ensure wherever possible that their money goes towards projects that benefit both the climate and the world's poor.

SULPHUROUS SUNSHADE

Taking aerosols as a whole, the world was at its dirtiest from the 1950s to the 1970s. By reducing incoming sunshine, sulphur, carbon and other particles in the atmosphere caused 'global dimming', which significantly cooled the Earth's climate. Some cold winters during these decades were even frigid enough to convince early climatologists that a new ice age might be on its way.[38] But as the West became wealthier, its inhabitants – forming the nascent environmental movement in the early 1970s – began to demand that acid rain, smog and other forms of air pollution be controlled. In the US, regulation culminated in the 1990 Clean Air Act amendments, which set up a successful 'cap and trade' programme to reduce sulphur emissions from American power stations. In Europe, progressively tighter regulation in the 1980s and 1990s has now reduced sulphur emissions by two thirds, and in some advanced countries by more than 80 per cent since 1990.[39] Humanity's progress in reducing these aerosols can even be tracked by looking at the moon, where 'earthshine' shows up more brightly when our atmosphere is more polluted and therefore more reflective.[40]

Aerosol pollution, like many environmental problems, is being solved not by reductions in consumption but by technology. In common with water pollution, air pollution tends to reduce as economies grow and people get more prosperous. There is no doubt that the benefits of cleaner air are many, including the reduction in acid rain, improvements in human health, and the reduction also in the negative climatic effects of aerosol pollution like the extended droughts suffered by North Africa in the last century. But one effect of aerosol pollution was undoubtedly positive, in that it held off global warming for several decades after the Second World War. Once pollution controls began to bite, however, temperatures began inexorably to rise in response: between 1985 and 2002 temperatures over land rose ten times as fast as during the previous two decades.[41]

All of this is evidence of climatic management by humanity, but to date this intervention has been conducted unawares as a consequence of other activities. The central argument of this book, however, is that humanity is today powerful enough, and increasingly knowledgeable enough, to begin to take a more intelligent approach. It is now an entirely serious proposition that we should shift the sunshade provided by sulphur aerosols from the lower troposphere, where sulphur dioxide and other suspended particles damage the health of millions and cause acid rain, to the higher stratosphere, where sulphates can continue to cut incoming solar radiation without anyone having to breathe them in. The idea is called 'solar-radiation management', and is one of many possible approaches for what is generally termed 'geoengineering'. In all likelihood, it is now technically and economically entirely feasible. But whether it is socially or politically desirable is another matter entirely.

One of the earliest and highest-profile proponents of solar-radiation management as a climate-change mitigation strategy was Paul

Crutzen, the Nobel Prize-winning atmospheric chemist and a member of the planetary boundaries expert group. Professor Crutzen, whose eminence as a scientific authority about the atmosphere is unimpeachable, pointed out in a landmark essay in 2006 that if just 2–4 per cent of the 55 million tonnes of sulphur humanity releases annually in the lower atmosphere were injected into the stratosphere instead, ongoing global warming could be substantially reduced.[42] Although sulphur particles last only for a week or so in the active weather of the troposphere, once they are floating safely above even the highest stormclouds in the stratosphere their atmospheric lifetime extends to a year or more. We know the idea would work, because nature does it from time to time through big volcanic eruptions. The last major eruption, of Mount Pinatubo in the Philippines in 1991, cooled the planet by half a degree for over a year.

Intentionally altering the Earth's albedo through geoengineering would be an enormous step for humanity to take. Consequently, the idea is vociferously opposed by many scientists as well as environmentalists. They have good cause for concern. The injection of aerosols into the stratosphere would likely reduce the strength of the African and Asian summer monsoons, potentially affecting water and food supplies for 2 billion people.[43] The 1991 Pinatubo eruption led to dramatically reduced rainfall and runoff throughout the subtropics the following year. 'The central concern with geoengineering fixes to global warming is that the cure could be worse than the disease,' warned climatologists Kevin Trenberth and Aiguo Dai following Crutzen's essay.[44] There is also the danger of 'moral hazard'. This is the idea that if a technological fix for global warming were made available, then governments would avoid the more difficult challenge of reducing greenhouse gas emissions. 'It's like a junkie figuring out new ways of stealing from his children,' was the angry response of Meinrat Andreae, an atmospheric scientist at the Max Planck Institute in Germany.[45]

The problem with these very legitimate objections – and there are many more too numerous to list here[46] – is that they fail to address the central conundrum identified by Crutzen: that by removing the aerosol pollution sunshade that currently reduces global warming by 50 per cent or more, humanity will expose itself and the planet to the full glare of the sun and consequent soaring temperatures.[47] Keeping our cities smoggy is not an option, for the reasons already outlined. So why not take up Crutzen's suggestion and simply move the sulphur sunshade higher up in the atmosphere? This challenge is one of the most critical questions facing humanity today, because it forces us to confront the unavoidable necessity of having to manage the planet intelligently. In my view, the arguments against intentional geoengineering are strong, but run the risk of amounting to a demand that we should merely stick to inadvertent geoengineering – that we should, in other words, go ahead and remove the aerosol sunshade anyway, and then just see what happens. I cannot readily accept that accidental planetary management is necessarily better than deliberate planetary management, so I think it is premature to reject geoengineering as a short-term and limited climate mitigation option.

This is particularly the case given the likely cheapness and technological feasibility of adding sulphates to the stratosphere. Crutzen estimates a global annual cost of $25–50 billion, less than a twentieth of the amount spent around the world each year on defence. He suggests high-altitude balloons or artillery shells as a way of delivering the necessary sulphur budget into the stratosphere, but there may be easier methods. Simply adding some sulphur-release mechanism to the world's 14,000-strong fleet of passenger aeroplanes, a significant proportion of which are likely to be flying in the stratosphere at any one time, would surely be a much easier task, requiring little additional effort. In addition, there is a whole variety of potential solar-radiation management technologies that could be deployed were stratospheric sulphur to fail or to misfire in some way. These include the idea of whitening low-level clouds over the sea by spraying tiny water droplets into the air, putting a few million small sunshades in space or even painting everyone's roofs white. Even the air capture of carbon dioxide by mechanical means, and its disposal underground, is considered by some to be geoengineering. In my view simply knowing what we are doing means that none of our actions in future that affect the climate can be called unwitting. Our hands are on the thermostat whether we like it or not, so sooner or later we are going to have face up to the need to make a decision about what temperature we want our planet to be at over the longer term.

I entirely accept, however, that there is a huge qualitative difference between an intentional and an unintentional action. No legal system gives the same penalty for premeditated murder as for accidental manslaughter. Nor would one expect the same degree of praise for saving someone's life by accident as for doing so on purpose. Accordingly, any decision on the deployment of geoengineering would need a step-change in our own systems of international policymaking and governance, a theme I will take further in the final chapter of this book. To repeat: no one could sensibly argue that solar-radiation management in any way negates the need for urgent and decisive action to reduce greenhouse gas emissions. We must reduce CO_2 emissions to zero anyway over the next few decades, not just to avoid catastrophic global warming, but also to prevent serious acidification of the oceans – another planetary boundary, examined in the next chapter.

But scientific research and development of different geoengineering strategies should be stepped up, and funding from world governments increased. I have no doubt that even flawed solar-radiation management with serious unintended outcomes would be much better for the planet and humanity in general than four, five or even six degrees of global warming. Geoengineering could be an important insurance policy against this nightmare ever happening. If tipping points are crossed that appear to be driving the climate system out of our control, we will have little option but to try to cool it down. For me this is a reason for optimism, and why I believe that my children, grandchildren and their contemporaries will never have to see the worst-case-scenario global warming I wrote about at the end of my last book, *Six Degrees*, because – barring some unforeseen worldwide civilisational collapse – humanity will have developed the technologies needed to avoid the holocaust of runaway global warming. Of course, curing the disease is always better than merely treating the symptoms. But doing both may be the best option of all.

CHAPTER NINE

The Ocean Acidification Boundary

I have already said a lot about human interference with the carbon cycle, because of the supreme importance of the climate change planetary boundary. But climate change has an evil twin, whose very existence was barely noted until comparatively recently, but which is now considered by the planetary boundaries expert group to be sufficiently critical to the Earth system to deserve separate consideration. This new boundary is the acidification of the world's oceans, which, as they absorb the carbon dioxide released by human burning of fossil fuels, are gradually turning more hostile to many forms of marine life.

Homo sapiens currently releases 10 billion tonnes of carbon per year – a million tonnes every hour. Since James Watt's invention of the steam engine in 1784, humans have released more than half a trillion tonnes of carbon from geological safe storage underground into the atmosphere.[1] Up to 85 per cent of this liberated carbon, somewhere between 340 and 420 billion tonnes, has soaked into the oceans.[2] This is a stroke of luck for us, because rates of greenhouse warming are sharply reduced as a result: were the oceans not performing this free service, the Earth's temperature would be rising at double or triple today's rate. But it is a service the oceans perform at a substantial cost to themselves, for by holding on to dissolved carbon dioxide they begin to change their chemical composition. The process is straightforward, involving what US marine biologist Richard Feely calls 'irrefutable chemistry'. To see the process in action, drop a piece of chalk into a glass of carbonated water. Why does the chalk dissolve in an angry froth of bubbles? Because carbon dioxide forms carbonic acid when it dissolves in water, and this acidic solution attacks the alkaline calcium carbonate that makes up chalk. The chemical equation is simple: $CO_2 + H_2O = H_2CO_3$ (carbonic acid).

The result is that, everything else remaining equal, the extra carbonic acid depletes seawater of the dissolved carbonate minerals that many marine creatures, from corals to plankton to sea urchins, use to build their shells or skeletons.[3] This 'other CO_2 problem' is now considered so crucial that one group of experts suggests ocean acidification 'could represent an equal (or perhaps even greater) threat to the biology of our planet' than climate change alone.[4] What the future holds is uncertain, as this chapter will show. As with climate change, we can base predictions on the educated guesses provided by computer models, and also peer into the Earth's deep geological past. My conclusion is straightforward. Even if there were no climate change, we would still have to get rid of CO_2 – urgently – because ocean acidification presents a serious threat to the integrity of the marine biosphere.

INDUSTRIAL OCEANS

At the dawn of the Anthropocene, at the end of the eighteenth century, the pre-industrial world oceans had a pH of about 8.2. Today ocean pH has dropped to 8.1 units and continues to fall. This may not seem like a big deal, but the pH scale (like the Richter scale for earthquakes) is logarithmic, so this small-sounding tenth-of-a-unit change translates into a 30 per cent rise in acidity in our seas.[5] As with the other planetary boundaries, we now have plenty of scientific data confirming the nature and magnitude of this change. The gradual acidification of the oceans is being constantly monitored at many different locations: at Station ALOHA in Hawaii researchers have observed a decreasing pH trend of 0.0019 units per year, precisely tracking the rise in atmospheric CO_2 measured since the 1950s not far away on the summit of the Mauna Loa.[6] Equal trends of rising acidity have been measured off the Canary Islands and Bermuda.[7] Whilst ocean pH fluctuates naturally, just like the weather, the ongoing trend towards acidification is as clear as that towards global warming. There is no significant doubt about what is happening, or the cause.

Again as with global warming, acidification 'hotspots' have begun to emerge at discrete locations in the world's oceans. Particularly vulnerable are west-facing continental coastlines, where water that is already naturally more acidic wells up from the depths. One group of scientists has discovered that 'corrosive acidified water' has reached 50–100 metres closer to the surface off the western United States, and that the trend is worsening rapidly.[8] Because colder water dissolves more CO_2 than warmer oceans, another acidification hotspot occurs in the Arctic, where melting sea ice in addition leaves more of the sea surface in contact with the atmosphere and therefore able to absorb carbon dioxide today than previously. In the summer of 2008 researchers on board the Canadian Coast Guard's heavy icebreaker the *Louis S. St-Laurent* found rapidly acidifying waters in the Canada Basin, an area of frigid seas north of the Canadian and Alaskan coasts. By measuring the salinity and chemical properties of the seawater they sampled, the scientists were able to conclude that this rising acidity was 'a direct consequence of the recent extensive melting of sea ice' in the area.[9] This suggests that the climate change and ocean acidification boundaries are linked not only in their cause, but also in their effects.

LIFE IN THE ACID BATH

Marine biologists are today some of the most worried people on the planet. Coral reefs, probably the world's most important oceanic habitat, are already in decline almost everywhere because of global warming, overfishing and other human impacts. But ocean acidification, unless rapidly addressed, will kill them completely. Reefs are entirely made of calcium carbonate, and that makes them extremely vulnerable to more acidic oceans. Past a certain tipping point, reefs will not just die – they will begin to physically dissolve in the corrosive water around them, and eventually disappear completely.

Already today there is evidence that acidification is beginning to have an effect on corals, even in protected areas like Australia's Great Barrier Reef. In January 2009 Australian marine biologists reported a 14 per cent decline in calcification rates measured over 69 different locations up and down the Barrier Reef over the fifteen years between 1990 and 2005. Because corals, like trees, lay down growth rings, the scientists were also able to note that these changes were unprecedented within at least 400 years – ruling out natural oceanic fluctuations and other recurring cycles.[10] Similar multi-year declines in coral growth have also been observed in the Caribbean Sea[11] and the Andaman Sea off Thailand,[12] likely also attributable to acidification, higher temperatures or both. Once again, the interaction between the climate and acidification boundaries is evident: laboratory studies with corals in tanks have shown that whilst hotter and more acidic waters are both bad for reefs, the combination of the two is particularly deadly.[13]

A natural laboratory for the impact of rising oceanic acidity can be found off Italy's western Mediterranean coast, where volcanic CO_2 bubbles out of vents in the sea floor. Scientists scuba-diving around these vents have noted that calcareous corals are unable to survive in the area, whilst other animals with chalky skeletons – like sea snails – show damage to their shells.[14] In the eastern tropical Pacific, scientists have found that coral reefs in more naturally more acidic seas are degraded and unstable.[15] The outlook is grim: even though some specific coral species seem to be resilient to acidification,[16] tropical reefs globally may start dissolving as early as mid-century if atmospheric carbon dioxide concentrations are allowed to double by that time from their pre-industrial level.[17]

The entire marine ecosystem is at risk from acidification, from the tropics to the poles. In the Arctic, bottom-dwelling organisms like mussels and clams – vital food for diving seabirds, bearded seals, grey whales and walrus – may already be suffering the impacts of rising acidity.[18] The acidification hotspot on the western US continental shelf could put at risk the giant kelp forests that are important centres of marine biodiversity, and affect lobsters, sea urchins, snails, mussels and hundreds of other ocean-dwelling animals and plants.[19] This and other upwelling zones that are experiencing the earliest effects of ocean acidification also tend to support flourishing fisheries – from the Humboldt current off the coast of Peru to Alaska's billion-dollar pollock fishery. Once again, the planetary boundaries interact in a worrying way: further research suggests that rising levels of acidity are likely to increase the area of low-oxygen 'dead zones', already affecting coastal areas around the world because of excess nitrogen and other waterborne pollutants.[20]

The impacts could be most severe at the base of the food chain, on which all higher life-forms depend. Although so small that they are individually invisible, plankton called foraminifera are present in such huge numbers in the world's seas that they account for between a quarter and a half of all the carbonate precipitated in the marine ecosystem – a much higher proportion even than coral reefs. Yet there is evidence that the tiny shells precipitated by these ubiquitous plankton are a third smaller than in pre-industrial times, perhaps in response to rising oceanic CO_2.[21] Other micro-sized planktonic calcifiers are the coccolithophores, which are so small – and so abundant – that they can be present in the tens or even hundreds of thousands in just a litre of seawater. By scattering sunlight through the upper layers of the ocean, they make it more opaque, and help give tropical seas their famed turquoise colour.[22] If acidification reduces their numbers, the oceans of our planet will subtly change colour – to a darker hue, perhaps with a greenish tint where photosynthesising algae take over from their calcifying brethren; another planetary-scale visual change that might be detected from space.

Scientific studies seem to confirm that coccolithophores are sensitive to ocean acidification: those organisms exposed to high-CO_2 conditions grow degraded or malformed shells, and reduce their rate of calcium carbonate production.[23] Similar effects have been observed for tiny planktonic floating snails called pteropods, which flourish by the tens of thousands per cubic metre of water in the Arctic and Antarctic seas, and are crucial prey food for pollock, cod, salmon and mackerel.[24] In the southern oceans, krill – important there as food for whales, seals, penguins and fish – are particularly vulnerable because their unique life-cycle means they migrate between the higher and lower depths of the ocean, and are exposed to very different levels of acidity in the process.[25] Acidification also changes the characteristics of seawater in unexpected ways. Acoustically, sound will travel further, changing the environment for whales and other cetaceans that communicate by hearing.[26] Some fish also seem to lose their ability to 'smell' properly in more acidic waters, even if they are not directly affected physiologically.[27] Some animals seem unaffected as adults, yet are vulnerable in their larval stage.[28]

But, again as with nitrogen, higher carbon dioxide will also have a fertilising effect, benefiting some opportunistic, weedy species. Seagrasses are expected to do well in a higher-CO_2 ocean. Although acidification will certainly reduce biodiversity overall, it will be a welcome boon for the seaweeds and choking algae that will coat the world's declining coral reefs as they erode and decay in the carbon-rich water.[29] Tiny organisms called diazotrophs, which can fix nitrogen in the sea just as leguminous plants do on land, also seem to proliferate as carbon dioxide levels rise.[30] If CO_2 emissions continue unchecked, the oceans will see a general shift away from calcifying organisms towards green algae and other photosynthesisers: a change characterised by marine biologist Jeremy Jackson as 'the rise of slime'.[31] As we saw in the biodiversity chapter, these kinds of shifts in species patterns can be expected to cascade up the food chain, where negative impacts on tiny plankton eventually affect food supplies for the great whales and other top predators. Accordingly, addressing this problem means understanding the interactions between at least four planetary boundaries: those on climate, nitrogen, biodiversity and, of course, ocean acidification itself.

REEF GAPS

As I showed in the chapter on climate change, scientists have learned a lot about the future by looking at the past. Over many millions of years, the Earth has seen periods of extreme global warming and associated intervals of very high atmospheric CO_2. Did the oceans turn more acidic then? If so, how did coral reefs and other oceanic life-forms survive through to the present day? Here we must learn from the deep-time insights of geology, and examine the fossil record for inklings of past events that might serve as analogues for the current human carbon release.

The major markers in geological time are the mass extinctions, of which there have been five major episodes in the last half-billion years. Corals, it turns out, were hit hard in all of them. At the end of the Ordovician period, about 434 million years ago, living reefs disappeared and did not reappear in the fossil record for 4 to 6 million years. Another 'reef gap' can be seen after the Late Devonian mass extinction, 360 million years ago. Yet another appears after the biggest mass extinction of all, at the end of the Permian, 251 million years ago. In that case, reef-builders disappeared for 10 million years, the biggest hiatus in all the Earth's history. When reefs eventually reappeared in the early Triassic, the old tubular corals had gone for ever, to be replaced by the scleractinian – or stony – corals that still dominate reefs today. The end of the Triassic saw another extinction pulse and reef gap, whilst the extinction that killed off the dinosaurs also exterminated the majority of corals.[32]

So was there a common cause? Most of the major mass extinctions correlate well with periods of sustained volcanism, when vast basalt 'provinces' flooded out from deep within the Earth's mantle, releasing colossal quantities of CO_2 in the process. The end-Triassic extinction, 200 million years ago, wiped out as much as half of life on Earth – and took place at the same time as an enormous volcanic episode tore apart the supercontinent of Pangaea, laying down a 7-million-square-kilometre area of 'flood basalts' that can today be found as far apart as the eastern USA, Brazil and Morocco.[33] Termed the Central Atlantic Magmatic Province by geologists, as this lava flooded out of cracks in the Earth's crust it simultaneously released between 8,000 and 9,000 billion tonnes of carbon in the form of CO_2, which in turn may have triggered another 5 trillion tonnes of carbon to belch out from the seafloor in thawing deposits of methane hydrate.[34] Even spread over ten or twenty thousand years, that is an awful lot of carbon – and the ensuing extreme greenhouse effect and bout of ocean acidification are recorded in both carbon isotopes[35] and a worldwide 'calcification crisis' as calcium carbonate-forming plankton were wiped out. In their stead, the hot and acidic oceans bloomed with green 'disaster' plankton species, whose remains depleted oxygen levels in the waters below and left them stagnant and anoxic.[36]

Fifty million years earlier, at the end of the Permian period, the biggest mass extinction of all time wiped out as many as 95 per cent of species – and once again volcanic CO_2, followed by a massive oceanic methane hydrate release, has been fingered as the main cause. All coral reefs, and most calcifying organisms, simply ceased to exist. This time the 'flood basalt' eruption took place in modern-day Siberia, and was also the largest of all time, releasing perhaps as much as 30 trillion tonnes of carbon. The rate of release has been estimated at around 1–2 billion tonnes per year for around 50,000 years.[37] Although the total carbon release was much greater than anything humans are likely to manage, the all-important rate of release was twenty times *slower* than modern fossil fuel burning. Even so, the ensuing reef gap in the fossil record lasts for 10 million years. Where corals did survive as organisms, they may have done so in 'naked' form in refuges, without their carbonate skeletons;[38] there is also evidence that calcium carbonate rocks were dissolving in more acidic ocean waters.[39]

A less ancient carbon spike – taking place a mere 55 million years ago – is perhaps the best analogue for the oceanic consequences of human fossil fuel combustion. The Palaeocene–Eocene Thermal Maximum (PETM) saw a rate of carbon release that was likely ten times slower than we are currently managing,[40] but overall a hefty 2–7 trillion tonnes of carbon were vented by a combination of volcanic activity, methane hydrates and peat oxidation, leading to a global warming of five or more degrees.[41] (Humans have so far released about half a trillion tonnes of carbon.[42]) Although in terms of its overall impact the PETM is a minnow among mass extinctions, it did lead to one of the top three worst reef-gap events in the entire Phanerozoic (the last half-billion years). There is clear evidence of global ocean acidification, with calcifying corals extinguished and green algae proliferating.[43] Indeed, the extinction of corals was close in magnitude to that of the much more dramatic end-Cretaceous extinction that killed off the dinosaurs.

So the geological record provides a clear and unambiguous warning that intervals of very high CO_2 release and global warming can cause dramatic ocean acidification and disaster for coral reefs. But this is not always the case. A hundred million years ago, during the mid-Cretaceous, chalky plankton thrived – despite atmospheric CO_2 reaching 2,000 ppm, and the pH of the oceans dropping as low as 0.8 units below today's levels.[44] These represent far more extreme 'hothouse' conditions even than the worst-case scenarios projected for 2100, and yet the thick chalk beds of southern England and elsewhere testify to the fact that these more acidic (but still alkali) oceans were extremely friendly to calcium carbonate-fixing organisms. There are other 'hothouse' episodes of extreme greenhouse conditions at different times in the Jurassic, Triassic and other eras that similarly left reefs seemingly undamaged.

The explanation for the apparent contradiction, however, is simple: time. The oceans are well buffered: as relatively acidic surface waters mix with those deeper down, sediments on the seafloor neutralise the acid – but only over tens of thousands of years.[45] Evolution can also work its magic over many millennia, allowing life-forms to adapt to a changed environment – and new species to emerge – through the ever-present pressure of natural selection. But in today's looming acidification crisis, time is not on our side. Having examined the published geological evidence of major carbon releases over the last 500 million years, I cannot find any episode – even during the darkest days of the worst mass extinction – that comes close to the current rate of human carbon emissions. In other words, all of our power-station chimneys, car exhausts, jet engines and so on across the whole world are releasing carbon dioxide an order of magnitude more rapidly even than the greatest super-volcanic eruption of the last half-billion years. I know this seems like an unlikely assertion, but then humans are an unlikely species. To get a sense of the scale of the enterprise, visit the website trillionthtonne.org, which constantly counts down the tonnes as they are released. The digits speed by so rapidly as to be unreadable; today's rate of emissions is about two hundred tonnes every second.

THE OCEANS OF THE FUTURE

The path forward is rather colourfully illustrated by one of the models. This projection of the future shows a map of the global oceans, assigning a colour gradient to the seawater saturation state of aragonite, the form of calcium carbonate used by coral polyps. Where the water is well saturated (blue), corals can flourish, but once acidification makes it less saturated (yellow and then red), reefs will not only stop forming but eventually begin to dissolve. The illustration shows six maps, each representing a different concentration of CO_2 in the atmosphere. Under pre-industrial levels of around 280 ppm, everything looks good for corals, with large areas of healthy blue throughout the tropical seas. In today's map (380 ppm), the blues are fading somewhat, with yellows creeping in from the edges as aragonite saturation drops. At 450 ppm the situation deteriorates further, whilst by 500 ppm almost all the blue has gone. By 550 ppm, the predominant colours are yellow and red, whilst in the final map – representing 650 ppm – reds and oranges throughout the entire ocean show that corrosive, acidified waters have now spread worldwide.[46]

Time is running out. The world's oceans are already more acidic than has probably been the case in at least 20 million years.[47] As early at 2016, five years from now, a tenth of the Arctic Ocean could be undersaturated with aragonite for at least one month in the year, making these waters toxic to calcareous organisms. By 2023, when CO_2 will reach 428 ppm if current emissions trends continue, 10 per cent of the Arctic Ocean will be undersaturated all year round, according to one 2009 modelling study, whilst by 2050, with CO_2 at 534 ppm, this area extends to half the entire ocean.[48] Accordingly, at a meeting of marine biologists at the Royal Society in London in July 2009 I was struck by the near-apocalyptic tone of the ensuing scientific paper. Noting that the crisis in the oceans adds to concerns that 'anthropogenic CO_2 emissions could trigger the Earth's sixth mass extinction', the experts warn that as early as 2030 'reefs will be in rapid and terminal decline world-wide', whilst by the latter part of the century they could be reduced to 'eroding geological structures with populations of surviving biota restricted to refuges'.[49]

Adhering to the proposed planetary boundary on ocean acidification is therefore critically important to the survival of the marine biosphere. The planetary boundaries expert group accordingly makes a very specific numerical recommendation, setting a maximum level of acidification that would allow coral reefs, plankton and other marine calcifiers to continue to exist. The crucial number regards the saturation state of aragonite, the most soluble form of pure calcium carbonate, and the one that is used by reef-forming corals and many other species to build their shells. A reasonable margin of safety, the planetary boundaries expert group proposes, would be keeping the aragonite saturation state at 80 per cent of pre-industrial levels, which should be sufficient 'to keep high-latitude surface waters above aragonite undersaturation and to ensure adequate conditions for most coral systems'.[50]

Given that aragonite saturation in the pre-industrial ocean was 3.44 and has now fallen to 2.9, we are currently at 85 per cent of pre-industrial levels.[51] The boundary is approaching rapidly, in other words, but we are still on the right side of it. Meeting it requires humanity to also respect the climate change planetary boundary, of reducing carbon dioxide concentrations in the atmosphere back below 350 parts per million, which in turn requires a carbon-neutral globe by 2050 and measures to extract CO_2 from the atmosphere thereafter. This is good news in that meeting the climate change boundary, which we must do anyway, will also protect the oceans from the threat of acidification. Here is another unassailable reason why we must act rapidly to decarbonise the global economy. Doing so will be a colossal challenge, but it is one we can successfully rise to, as I outlined in Chapter 2, by combining the best of modern zero-carbon energy technology with a more creative approach to raising the large sums of capital financing that are necessary to fund the transition away from fossil fuels.

There are some short-term palliative measures that can possibly be taken, though these are no substitute for the central challenge of reducing carbon emissions. One idea is for the use of ground-up olivine rock to be spread as beach nourishment in areas where beach sand is disappearing. Available in immense deposits around the world, olivine gradually dissolves in seawater to create bicarbonate ions out of carbonic acid, helping to fertilise corals at the same time as reducing ocean acidification. If used on a large enough scale, waters around the world's largest coral reefs could be kept in a more alkaline state by the constant addition of olivine, researchers suggest, and the process would have the additional benefit of drawing down carbon dioxide permanently from the atmosphere.[52] In effect, this is simply mimicking the natural weathering process of rocks that stabilises the carbon cycle over geological timescales, but in a dramatically accelerated form. Given the need to mine, crush and ship olivine rock, the cost is hard to judge, but it may be low enough to be a commercially competitive form of carbon sequestration, as well as a mitigation strategy for ocean acidification. At this stage, the urgent need is for further study, and trialling on a small scale to test impacts on the marine environment.

Another possibility is the addition of lime directly into the oceans. The production of lime out of limestone uses energy and releases CO_2, but once the resulting powder is added to the oceans almost twice as much carbon dioxide is absorbed, according to the idea's promoter, Tim Kruger. If concentrated around coral reefs, the process could directly address acidification. By sequestering CO_2, it would also be carbon-negative. Kruger estimates that one and a half cubic kilometres of limestone would be needed to draw down a billion tonnes of carbon from the atmosphere.[53] Large-scale desert solar thermal plants or nuclear reactors could be used to generate the initial heat energy in a zero-carbon way. As with the olivine idea, I would not see this as an alternative to emissions reduction, but humanity will need to develop technical solutions for removing the additional CO_2 that has already accumulated in the atmosphere above the climate change planetary boundary of 350 ppm. No one is suggesting immediate deployment, but carbon-negative solutions that also directly address ocean acidification at a chemical level are surely worthy of additional research and development in order to properly assess their real-world feasibility.

Overall, there needs to be a step-change in the level of attention given to the issue of ocean acidification. So far, only academic conferences have addressed the challenge properly – public knowledge and concern remains extremely low. But, urgent as the situation is, scientists, environmentalists and others concerned with raising awareness must act carefully, and ensure that they do not repeat the mistakes that have stymied global progress on climate change. Appeals to public action must be framed positively, and not fall into the trap of sparking a media and political backlash. Sacrifice and austerity are out; competition and innovation are in. The 'deniers' are waiting in the wings, and if the wrong triggers are pressed it will only be a matter of time before any constructive debate on mitigating ocean acidification is destroyed by the same politicised trench warfare that has overtaken the debate on climate change.

SCEPTIC TANK

Ocean acidification sceptics are still few and far between, but they do exist. Almost exclusively, they take their cue from the climate debate, simply extending their contrarian stance on global warming into this newly unfolding scientific arena. This straightforward intellectual extrapolation is stated quite explicitly by the British writer Matt Ridley, who writes provocatively in his 2010 book *The Rational Optimist* that 'ocean acidification looks suspiciously like a back-up plan by the environmental pressure groups in case the climate fails to warm'.[54] The case against ocean acidification as a valid environmental concern has accordingly been taken up by a growing number of climate-denialist groups and websites, which label it variously as 'the next big hoax' or a 'scam'. The London-based climate-sceptic think-tank the Global Warming Policy Foundation carries frequent contrarian articles on the issue on its website, whilst luminaries who have made a career out of denying the reality of global warming are now beginning to converge on ocean acidification as being, in the words of the Cato Institute's Patrick Michaels, 'the next hysteria'.[55]

In scientific terms, the criticisms levelled by Ridley and other ocean acidification sceptics are no more valid than the reasons they have already given for not believing in global warming. In a November 2010 op-ed for the *Times* newspaper,[56] Ridley draws an analogy with acid rain, which he asserts was also the 'scare' of its day. Utilising all the classic tactics of climate-change deniers, he uses this false analogy (it is false because acid rain was real) to back a charge of grant-funding 'vested interests' by marine scientists, before cherry-picking (and misinterpreting) a small number of scientific studies that appear to contradict the consensus position on the biological impacts of ocean acidification, and generalising on this basis that the whole thing is exaggerated. He ties this to a number of mistakes, writing, for example: 'Environmentalists like to call this [end of the century projection] a 30 per cent increase in acidity, because it sounds more scary than a 0.3 point decrease (out of 14) in alkalinity ...' Actually, if we recall that the pH scale is logarithmic, a projected 0.3 point decrease in pH represents a doubling (100 per cent) in ocean acidity, as a response statement by a group of marine scientists points out.[57] The error is particularly ironic because Ridley is lambasting 'environmentalists' for apparently using numbers that underestimate the true scale of the acidification problem.

Note in particular the recurring conflation of 'scientists' with 'environmentalists': Ridley's narrative depends upon convincing the reader that there is no serious difference between the two, and that therefore peer-reviewed science published by eminent researchers is no more valid than a Greenpeace press release – and perhaps just as self-interested. He begins the article indeed by deliberately aligning the acidification issue with environmentalism: 'As opinion polls reveal that global warming is losing traction on the public imagination, environmental pressure groups have been cranking the engine on this "other carbon dioxide problem"'. Ridley rounds off the piece by adopting the voice of the unbiased ordinary person and lambasting 'the media' (always a convenient straw man) for allegedly not reporting the truth. 'Before I started looking into this, I assumed the evidence for damage from ocean acidification must be strong because that is what the media kept saying. I am amazed by what I have found,' he concludes.

This is strong stuff. But what interests me most is not the scientific content of Ridley's accusations so much as the political motivations that clearly underlie them. Ridley is undeniably passionate in his belief. I make no accusation that he is in the pay of commercial vested interests – his opinions are obviously genuinely and deeply held. One could argue that Ridley's stance is actually a reputational risk for him, because in denying what scientists call the 'irrefutable chemistry' of ocean acidification (and adopting a similar position on climate change) he undermines his credibility as a popular science writer on other issues. This is curious, because *The Rational Optimist* is in many other subject areas a well-researched, illuminating and intelligent book, ranking with the best I have read for many years on the subjects of human physical and cultural evolution. So why take this risk with his reputation as a reliable communicator of scientific knowledge?

The answer, of course, lies in the political polarisation that has overtaken the societal debate about carbon emissions in particular and the environment in general over the last few years. Whilst ocean acidification does not yet register high enough in public awareness terms to be seriously at issue, opinion polls show very clearly that for climate change there is now a sharp left–right split amongst voters. Nowhere can this be seen more clearly than in the United States, where for the November 2010 mid-term elections the Republican Party fielded a slate overwhelmingly dominated by candidates questioning the science on climate change.[58] On the more extreme right represented by the Tea Party movement, climate denialism is an article of faith (and I mean that literally). In the UK, climate-change deniers I know (most of whom I have great respect for on a personal level) all seem to hail from the political right, Matt Ridley included. The Global Warming Policy Foundation was founded by Nigel Lawson, formerly Chancellor of the Exchequer in Margaret Thatcher's government, and author of a 2008 book called *An Appeal to Reason: A cool look at global warming*, whilst the redoubts of newspaper climate denialism are all in the right-wing press, from the *Daily Mail* to the *Telegraph* and the *Spectator*.

Why should this be? Are many right-wingers, as some on the Green left assert, simply 'anti-science'? Part of the explanation I think lies with deeply held political and social values. Climate-change deniers tend to be suspicious of what they see as big government and potentially intrusive state regulation over the private affairs of citizens and businesses. At the extreme represented by the US Tea Party, this extends into a wide-ranging conspiracy theory about government in general: its self-styled 'Declaration of Independence' talks of Democrats 'seeking to impose a Socialist agenda' and the 'expansion of government power' that can only lead to 'tyranny' and the reversal of capitalist progress.[59] I have met and debated with many climate-change contrarians over the years, and my overwhelming impression is that they are motivated more than anything else by a libertarian political agenda that rejects the assumed universalist ideology of environmentalism, and that most also maintain a strong faith in the power of free enterprise to overcome any social or environmental problem. The science, of course, comes second: people tend to select what they want to hear based on pre-existing beliefs, and choose their sources of information accordingly. A politically polarised media and the rise of blogs as a mainstream source of news and information have only reinforced this divergence in recent years.

Another issue under contention goes to the very heart of this book. I write here about fundamental planetary ecological limits, physically hard-wired boundaries in the Earth system that can be identified scientifically and that humans must learn to respect if the planetary system as a whole is to remain reliably stable and hospitable to our species and many others. Yet most enthusiasts for free markets and capitalist economics find the idea of ecological limits hard or impossible to accept, and choose to believe instead that the expansion of human material consumption – even on an obviously limited single planet – can continue for ever. As I will show in the final chapter, I believe that these diverging points of view can be reconciled, but only if both sides of the debate are prepared to be more open-minded. I have been frequently critical of the environmental movement in this book, but my argument with conventional environmentalism lies more with strategies than ultimate objectives: if the science about planetary boundaries is ever to gain popular credence and political force it is surely the Green movement that must be its most passionate and determined champion.

Climate change (and, by extension, ocean acidification) is politically toxic to the libertarian right precisely because it forces humanity to confront the necessity of respecting planetary limits – in this case regarding the capacity of the Earth system to tolerate emissions of greenhouse gases. I do not join with many Greens in accusing right-wing climate sceptics of being 'anti-science': as I have shown in earlier chapters, Greens are equally anti-science when it suits their own biases on issues like nuclear power and genetic engineering. But I do believe that the political right, with which I share some sympathies in many ways, has committed a serious error in cleaving to a dead-end denialist position on climate change – and thereby sacrificing the opportunity to frame economically liberal and free-market policies with which to combat it. At the same time, the capture of the Green movement by the political left has reinforced this sorry divergence, rendering Greens more marginal and less credible in the popular mind at the same time as reinforcing the political backlash from the right. Currently the Green left seems determined to dig itself still further into this political cul-de-sac, preferring to urge an unappealing narrative of communitarian austerity on an unwilling public. The insistence that people must give up cars, live in colder houses or holiday close to home is obviously a losing strategy, yet the flawed assumption that ecological limits like climate change mean that people must limit their own aspirations and lifestyles is central to mainstream Green thinking.

A good example of this dead-end ideology from the Green left side was the launch in January 2011 by the UK Green Party and New Economics Foundation of a report that adopts this hopeless strategy of austerity and sacrifice as a central part of its messaging. Called 'The New Home Front', the report insists that Britain must return to wartime policies of rationing and community solidarity to confront the modern-day 'emergency' of climate change.[60] Britain's only Green MP, Caroline Lucas, made the symbolism all too clear by holding the launch event in the Imperial War Museum. In its conclusion, the report argues: 'If we are to overcome the threat of climate change, our country will need to move onto the equivalent of a war footing, where the efforts of individuals, organisations and government are harnessed together and directed to a common goal. Only this will provide the urgency, energy and creativity we need to avert disaster.' This 'war footing' would seemingly involve an endless nationwide emergency, with climate-change propaganda plastering every billboard, big-screen TV owners and SUV drivers labelled 'antisocial', enforced rationing of carbon emissions throughout the entire population, and some kind of 'carbon army' patrolling the streets.

It would be easy to laugh this off were it not so obviously counter-productive. The idea that tackling climate change means accepting profound levels of intrusion into our everyday lives – and the economic disaster of dramatic drops in consumption and living standards – is an illusion that is actually shared by the Green left and the libertarian right: the former insist we must all submit to state-sponsored rationing, whilst the latter are so terrified of the prospect that they deny the very existence of climate change for fear of the political consequences they assume are inherent in addressing it. My argument here is that they are both wrong, most especially because the Greens do not – and should not – have a monopoly over legitimate policy responses to the carbon problem. In reality we *can* build our way out of climate change using a smart combination of innovation, investment and regulation, as I showed in Chapter 2. Lucas and NEF, fanatically anti-nuclear, leave themselves little option but to insist on dramatic cuts in energy use because they reject a main source of zero-carbon power. Reading through their proposals, I find myself agreeing instead with the responses of some right-wing pundits, who viewed the recommendations in 'The New Home Front' report as a recipe for economic disaster or worse.[61]

My hope is that if we can persuade influential climate-change sceptics – and thereby by extension the wider portion of the public which suspects that global warming is exaggerated – that the Green left's proposals for tackling climate change are not the only options on the table, then we can begin to move beyond the denialist backlash and start to craft a political narrative that places growth, innovation and aspiration at the centre of our response to the real challenge of ecological limits. If we can achieve this before the political battle lines are as firmly entrenched on ocean acidification as they are on climate change, then we may still have a chance of respecting the planetary boundary before time runs out and we cross over into the danger zone.

CHAPTER TEN

The Ozone Layer Boundary

Few things can better illustrate the awesome power of our species than the fact that a single human being can nearly destroy an important part of the Earth system almost singlehandedly. Thomas Midgley was one such person. An obscure American chemist, whose main inventions took place in the 1920s and 30s, this one man had 'more impact on the atmosphere than any other single organism in earth history', as historian John McNeill puts it.[1] This was not Midgley's intention, of course. As a loyal servant of General Motors Corporation, he merely had the task of seeking non-flammable coolants for the company's Frigidaire division, which was looking to replace the explosive and dangerous methyl chloride gas used in its fridges. Midgley's solution was Freon, his trade name for a non-flammable synthesis of chlorine, fluorine and carbon. Today we know it better as chlorofluorocarbon, or CFC.

The ozone-layer story demonstrates the double-edged sword of human technological prowess. On the one hand, modern science allows unremarkable individuals to have remarkable impacts at the planetary level through the deployment of everyday technologies that later turn out to be unintentionally but extremely damaging. On the other, science now equips us with the knowledge to at best identify these damaging impacts before they happen, or at worst to notice when they appear and take steps to counter them. Until recently, history suggests that technological applied science has tended to outstrip environmental science, as some of the other planetary boundary areas show. For example, the Haber-Bosch process to create synthetic nitrogen was invented early in the twentieth century, and was producing millions of tonnes of ammonia per year before the downsides of nitrate pollution and oceanic dead zones were spotted. (As John McNeill notes: 'Midgley was the Fritz Haber of the atmosphere.') Similarly, toxic pesticides like DDT were in very widespread use before their negative effects on bird species and biodiversity in general were noted.

Humanity was both lucky and clever when it came to the ozone layer. We were lucky because had Thomas Midgley made a different choice from the periodic table he habitually carried around in his pocket – using bromine rather than chlorine as the basis for his new synthetic Freon refrigerants – almost the entire planetary ozone layer would have been destroyed as early as the 1970s, well before the science of atmospheric chemistry was well enough developed to recognise the problem.[2] This is because bromine is far more destructive even than chlorine in the stable parts of the upper atmosphere where the ozone layer forms. But humans were also clever because our understanding of atmospheric chemistry did develop, in leaps and bounds in the 1970s and 1980s, and consequently when the real-world ozone hole did open up over Antarctica, scientists – who had already raised theoretical concerns about the impacts of CFCs – were quick to spot it and rapidly identified the likely cause.

Some of the key figures in the development of ozone-layer science are worth noting. The first person to accurately measure the gradual accumulation of CFCs in the atmosphere was none other than the Gaia hypothesis originator and intellectual father of Earth-system science, James Lovelock. On a research ship to Antarctica in 1971–2, Lovelock took the opportunity to test out his new invention, an electron capture detector for gas chromatography. This exquisitely sensitive instrument allowed for the first time the measurement of trace gases present in the air in only tiny quantities. On board Lovelock discovered that CFC (chemical formula CCL_3F) was present at around 50 parts per trillion everywhere he took measurements, from the green coasts of Ireland to the storm-lashed Southern Ocean. This was the first scientific proof that a wholly man-made gas had become ubiquitous throughout the planet's air. It was an epochal discovery, but far from raising the alarm, in his resulting scientific paper (published in *Nature* in 1973) Lovelock instead asserted reassuringly that 'the presence of these compounds constitutes no conceivable hazard' – a claim he later acknowledged as a 'gratuitous blunder'.[3]

Fortunately this error was quickly corrected. The most important leap was made just a year after Lovelock's report, in another *Nature* paper, this time written by Mario Molina and F. Sherwood (Sherry) Rowland – perhaps one of the most environmentally important scientific publications of all time, and one that was to gain the two men a Nobel Prize.[4] In June 1974 Molina and Rowland proposed that CFCs in the stratosphere could be split apart by ultraviolet light, leading to dangerously reactive free chlorine atoms circulating high in the atmosphere. These atoms could destroy stratospheric ozone on a scale that no natural process had so far achieved, seriously endangering life at the Earth's surface. Their work was taken further by Paul Crutzen, the Dutch-born atmospheric chemist who is today also a member of the planetary boundaries expert group (and mentioned in an earlier chapter for his proposals on geoengineering). Crutzen calculated that up to 40 per cent of ozone could be destroyed in the highest regions of the stratosphere by chlorine-containing CFCs, with the most important ozone-destroying reactions taking place in extremely cold, thin ice clouds high in the springtime polar stratosphere.

And so it proved: in 1985 scientists based at the British Antarctic Survey's Halley Base noticed that stratospheric ozone concentrations were plummeting in the frigid air of the Antarctic spring – with up to 40 per cent of the ozone layer depleted in a phenomenon quickly dubbed the 'ozone hole'. NASA, which had also spotted the hole but first assumed it to be an instrumental error, quickly confirmed the thesis with worldwide satellite measurements. Theory and observation aligned, and Midgley's CFCs were generally accepted as being the main culprits. Further damning evidence came when falling ozone levels were also identified in the Northern Hemisphere, raising the perils of skin cancer and eye cataracts for large populations in countries like the United States.

HUMANITY'S FINEST HOUR

The Vienna Convention for the Protection of the Ozone Layer, signed in 1985, is often cited as the most successful achievement ever in multilateral environmental policymaking. Actually the Convention was mostly symbolic: the real breakthrough came two years later in Montreal, with the signing of a Protocol amongst the major industrialised countries that began the phasing out of CFC production. The story is worth examining in detail, not only because it shows that humanity can successfully navigate away from a breach of a planetary boundary, but because studying the way in which policies were developed and agreed internationally can potentially help us tackle even thornier issues – starting with climate change.

There is no denying that the international ozone regulation regime has been startlingly successful. Humanity was over the planetary boundary – suggested by the expert group at 5 per cent loss of stratospheric ozone below its pre-1980 level – for just five years, between 1992 and 1998. We are now well within the ozone boundary: recovery has been slow, but losses currently average around 3 and 4 per cent, and the situation is expected to gradually improve.[5] The Antarctic ozone hole stopped growing a decade ago: its all-time record was set on 9 September 2000, at 29.9 million square kilometres.[6] Looking forward, experts project that the Earth's ozone layer will recover back to its pre-1980 level by sometime between 2060 and 2075.

In navigating back within the ozone boundary, humanity certainly avoided a hellish future. One NASA study suggests that without the Montreal Protocol two thirds of the Earth's ozone layer would have disappeared by 2065.[7] The ozone hole over Antarctica would be a year-round feature, whilst another hole would have opened up over the North Pole too. UV radiation falling on mid-latitude cities like Washington or London would be intense enough to cause sunburn in as little as five minutes. Ecological effects would have been devastating, and additional human skin cancer rates would be in the hundreds of thousands or even many millions. The ozone layer, essential for terrestrial life and a constant feature of the biosphere for 2 billion years, would have been largely destroyed by humanity in just a few decades.

So how did we manage to make such a good job of ozone regulation? First, the fallacies. There is a pernicious myth, widely believed in the environmental movement today, that tackling ozone depletion was in hindsight somehow 'easy', as if it was almost bound to happen. In fact, the negotiations were tortuous and extremely difficult. Whilst the basic science was established as early as the mid-1970s, it was to be another decade before there was any breakthrough in international policymaking. During this decade of paralysis, the CFC industry fought hard and successfully to avoid any international controls on the production of ozone-depleting substances, and powerful blocs of countries stymied the attempts of some more enlightened governments to move forward more quickly with the protection of the ozone layer.

Another related myth is that the CFC industry switched away from opposing regulation when it realised that easily available substitutes could be manufactured and profited from. A rather conspiratorial variant of this theory posits that the chemicals company DuPont, the largest of the CFC producers, changed its position when it secured a patent on an important CFC substitute. The truth is much more challenging. Although it was widely known quite early on that developing substitutes to CFCs was technically feasible, all the major companies – including DuPont – stopped their research programmes in the early 1980s. Why? Because they preferred to stall, and for as long as they believed regulation could be avoided the corporations decided it was better not to know about substitutes for fear that they would be forced to develop them. It was therefore generally believed by almost everyone, and frequently asserted by the companies, that developing substitutes for CFCs in the important areas of refrigeration, electronics and foam-blowing would be prohibitively expensive.[8]

This deadlock could have continued for decades more, as it has to a large extent with climate. The industry was well-organised and funded: in the US it formed an Alliance for Responsible CFC Policy, and ran campaigns targeting politicians and the media to emphasise its view that the science was uncertain and that ozone policy would be costly and economically damaging. Yet between 1985 and 1987 something dramatic changed, which turned the whole situation upside down. By the time the Montreal Protocol was agreed, DuPont had stopped attacking ozone science and announced its support for strong curbs on CFC emissions, as had the Alliance for Responsible CFC Policy. Instead of trying to stymie progress, industry had turned into a partner for change, and began an aggressive effort to develop and commercialise alternatives to CFCs: in 1988 DuPont announced that it would stop manufacturing CFCs altogether and instead supported a worldwide phase-out. There was no secret technological breakthrough: the company's stock value actually declined after the announcement. The new regulations that DuPont supported were putting a $130 billion industry at risk. So why the switch?

Part of the answer was science. The discovery of the ozone hole in 1985 helped focus public awareness, as did the publication of satellite data that seemed to show substantial ozone losses over much wider areas. Much more of a difference was made by the publication of authoritative scientific assessments by NASA and the World Meteorological Organisation which underlined the scientific consensus that further growth in CFC production would bring major damage to the ozone layer. These 'blue books' on the science helped convince vacillating policymakers that the ozone issue had to be tackled, cemented public support, and made a denialist industry position look environmentally irresponsible.

But there was more to it than science. Authoritative assessments were important because they could form a common point of reference, a shared standard for everyone – even industry. But another force was gathering, and that force was the one that made the real difference. This was political leadership. What happened with the Vienna Convention in 1985, and more importantly with the Montreal Protocol two years later, was that governments decided to make a leap of faith: even in the absence of conclusive evidence that substitutes for CFCs would ever become commercially available, they were prepared to regulate their production in the interests of the global environment. And sure enough, when CFC manufacturers realised that strong international regulation was likely or even inevitable, their financial incentive moved away from blocking progress towards developing alternatives that would position each company profitably in the new markets. There was no silver bullet: many varying substitutes to CFCs had to be developed for different industrial processes and applications, from the refrigerants in car air-conditioning systems to the solvents used in electronics manufacturing. Some were difficult and expensive at first, but industry innovated and costs inevitably fell.

So the politicians led, and private industry delivered in consequence. And as substitutes to ozone-damaging substances began to become available, so governments were emboldened at subsequent negotiations to tighten up the Montreal Protocol even further, and move towards a total worldwide ban on CFCs. There was a ratcheting upward effect, where confidence grew, new commercial incentives began to appear, and a steadily stronger ozone regulation regime came into being than was ever originally envisaged. Looking back, it now seems obvious that the original arguments made by industrial vested interests – that alternatives to CFCs were unavailable or too costly – were flat wrong. Not only did these technical alternatives appear more quickly even than the most fervent promoters had dared to hope, but in many cases they actually saved industry money. But at the time this was far from obvious. What happened was that the leap of faith taken by governments led to a tipping point, which turned the whole dynamic of the ozone issue from one of stasis to one of rapid change.

What is even more extraordinary is that the government that was most ambitious in leading worldwide action on the ozone layer is the long-time *bête noire* of climate-change negotiations, the United States. In contrast, European governments, and especially that of the UK, fought a long campaign against strong international restrictions on CFCs in order to protect their chemicals industries. They were forced to compromise, however, by the threat of unilateral Congressional action and American trade sanctions against their products. The United States made clear that without a global deal, it would enforce a de facto ban through sheer economic muscle. Even more strangely, all this took place during the anti-regulation administration of President Ronald Reagan. It was led by the Environmental Protection

Agency, but also by strong pressure within the United States Senate. Again, in stark contrast to the climate-change issue, the Senate enthusiastically ratified the Montreal Protocol in March 1988, making the United States only the second country to do so.

So what are the lessons for climate change and the other planetary boundaries? First for me is that pessimism leads nowhere. There were many reasons for pessimism even on the eve of the Montreal Protocol: world production of CFCs was still growing rapidly, and massive new investments in productive capacity were planned. Hundreds of billions of dollars of future profits were at stake, and powerful industries were vociferously opposed to meaningful change. Major developing countries were not on board either. In a striking parallel with carbon, China had announced plans to increase CFC production tenfold by 2000. And yet industrialised-country governments were prepared to take a leap in the dark, and a tipping point was thereby crossed. Within just a decade from the signing of the Montreal Protocol, worldwide CFC emissions had fallen by 95 per cent. Had the pessimists been listened to in 1985, I believe that the ozone layer would still be thinning today. Pessimism when it comes to political change can only ever be a self-fulfilling prophecy, and therefore has no place in policy.

The lesson ensuing from this is that nothing can change without strong political leadership. Industry cannot shift by itself: any company voluntarily eliminating a damaging product that is also made by others will simply cede ground to a competitor. Nor can the private sector be relied upon either to have an environmental conscience or to see the opportunities inherent in progressive change: vested interests in the commercial status quo will always be more powerful than potential winners not yet making profits in the new markets of the future. This does not mean that corporations should be demonised, however, for they may become vital partners in any eventual economic and technical transformation once governments have decided that such a change is necessary. But the tipping point towards action can only be crossed by politicians, and to do that the corporate lobbyists seeking to protect the status quo must be ignored in the wider interests of human society and the global environment.

HUMANITY'S DARKEST HOUR

For climate change, the necessary political leap of faith was almost made with the Kyoto Protocol, signed in 1997 and designed in explicit imitation of the Montreal Protocol of a decade earlier. But one crucial participant was lacking: the United States. Even before Kyoto, the US Senate, bowing to pressure from the fossil-fuels lobby, made clear that it would never ratify an international treaty on climate change. Represented by Vice-President Al Gore, the Clinton administration nevertheless took on a 7 per cent emissions reduction target under Kyoto, but never submitted it to the Senate. Once Clinton and Gore had lost the argument for Kyoto domestically, the stage was set for the new regime of President George W. Bush to repudiate the Protocol entirely. Not only would the US not lead on climate, it quickly grew clear, but it would do everything it could to thwart international progress. Without America, no other political bloc – not even the European Union – was strong enough to take the necessary leap of faith to convince the fossil-fuels industry that its days were numbered. It is a great tragedy, perhaps the greatest in the environmental field, that the United States blocked Kyoto. Had America backed action on climate as squarely as it tackled the ozone layer, we would be in a very different place today.

The Montreal Protocol was also much more successful in the way it dealt with developing countries. Although arguments were fierce over whether poorer countries should also have to eliminate CFC production, they were resolved by giving developing countries an extra decade to comply with the new regulations, and setting up a properly funded financial mechanism to help them do so. The issue was not dealt with at the initial Montreal meeting in 1987, but in subsequent negotiations that tightened up the ozone-regulation regime and brought in new players as confidence grew that the system would work. In contrast, the Kyoto Protocol set up a permanent 'Berlin Wall' between rich countries with emissions targets and poor countries without them – a deal that now looks rather anachronistic given the rapid rise of China, India, Brazil and other large new emitters in recent years. This arrangement has maintained a bitter ideological divide, where poor countries accuse rich ones of betraying their commitments, whilst rich countries worry that any emissions cuts they make will be overwhelmed by the growth of the poor. Both sides have some justice in their accusations, but their never-ending battle has brought the Kyoto Protocol to its knees, and led to an unsightly debacle in Copenhagen in 2009 that nearly destroyed the entire multilateral climate process.

Another difference between ozone and climate is that authoritative scientific assessments have not been as successful in convincing naysayers about the latter as they were with the former. This is not due to any shortcomings in the scientific process: evidence about the reality of global warming is far more overwhelming today than it was about the threat to the ozone layer in the mid-1980s. Nor have the experts failed to speak with one voice: the Intergovernmental Panel on Climate Change has delivered unimpeachably weighty assessments over the years, underlining its growing confidence about the science on climate change. But with climate the reactionary backlash has been unprecedentedly successful. It is almost forgotten now, but there was a denialist backlash against ozone regulation too, centred in the US in the mid-1990s, which swayed some important politicians. But the ozone deniers never made the leap into the centre of the political stage that the climate deniers have managed. Once again, I believe that America was a crucial force here: denialism on climate change has been aggressively promoted by a variety of right-wing think-tanks – many of them part-funded by the fossil fuels industry – whose ideologists in recent years have managed to position global-warming denial as a centrepiece ideological stance for the Republican Party. Copycat think-tanks and ideologues, again almost exclusively on the libertarian political right, quickly sprouted up in other countries too.

Indeed, climate denialists became so successful in 2009 that they managed to dominate the media agenda via a series of manufactured scandals that engulfed much the climate-science community. Deniers promoting the so-called 'Climategate' affair took a few out-of-context quotes and superficially embarrassing private slips by leading scientists from some leaked emails, and nearly managed to publicly discredit not only the Climatic Research Unit of the University of East Anglia but several other leading institutes too. Vociferous promoters of a subsequent scandal took a single mistake about Himalayan glaciers, buried deep in the second weighty tome of the IPCC's 2007 Fourth Assessment Report, and used it to attack the entire IPCC process, and the role of the Chair Rajendra Pachauri in particular. None of this changed anything we knew – anything that mattered – about the reality of climate change, but the deniers succeeded in making climate science an ideological battleground where the expert consensus was rejected by whole political parties and large sections of the media as itself partisan.

The failure of climate policymaking has been a self-reinforcing process. Industry carries much of the blame for entrenching itself in a position that for too long opposed political action. With the all-important political tipping point not crossed, companies that had begun to reposition for the post-carbon age instead began to fall back into their old roles. For example BP, which for a few years rebranded itself as 'Beyond Petroleum' and flirted with solar power, moved strongly back into fossil fuels, even investing heavily in oil extraction in the dirty Canadian tar sands. Of course, it is also important to recognise that moving out of fossil fuels was always going to be orders of magnitude harder than disavowing CFCs – oil, coal and gas provide most of the energy powering modern industrial civilisation, rather than being important for just a small number of specific uses and processes. The companies invested in fossil-fuel production are commensurately vastly more politically and economically powerful. Nor do oil and coal companies have much likely future in a clean-energy society. Whilst DuPont, for which CFCs were only a small section of its business anyway, could reposition itself to make a fortune selling CFC substitutes, it is difficult to imagine any existing fossil-fuels company making a serious contribution to decarbonising our economy.

But, as I outlined in the previous chapter, I also believe the environmental movement is partly responsible for this ongoing failure by promoting an alignment of climate-change mitigation with austerity, sacrifice and high cost. By analogy, had action on ozone required permanently eschewing hairspray – to say nothing of giving up refrigeration – the Montreal Protocol would likely be as stalled and ineffective as Kyoto is today. Accordingly, having missed our first chance for a political tipping point, we need to build momentum for a second chance by convincing people that substitutes for the energy services provided by fossil fuels are easy, cheap and effective.

I will discuss how regulating all the boundaries together might work internationally in the next and final chapter. But for now, let the lessons of the ozone layer be seen clearly in how they relate to climate change first and foremost. Most crucially, there is no need to wait for new energy technologies before forging strong international agreements – these technologies either already exist or will appear when they are needed. There are plenty of substitutes for carbon, but there is no substitute for political leadership. The pressure of regulation is essential to drive the technical innovation that will help us avoid the worst of global warming. And finally: the true prize is not incremental percentage cuts in emissions, but – as with CFCs – the elimination of fossil-fuels use altogether from the entire world economy. Nothing less will do the job.

CHAPTER ELEVEN

Managing the Planet

'All the great laws of society are laws of nature'
– Thomas Paine, *The Rights of Man*, 1791

THE COPENHAGEN DEBACLE

It was early December 2009, and I was stuck in a large conference hall on the outskirts of Denmark's capital city, Copenhagen. Outside it was snowing, and hundreds of people were queuing to gain admittance. Inside, tens of thousands more bustled around the central hall, the general hubbub punctuated now and then by the shrill cries of campaigners tirelessly advocating their various climate-related causes. As to the progress of the all-important negotiations, there was no news. At that moment, with just a few hours to go, it looked as if the 2009 Copenhagen climate-change conference, the much-hyped and much-heralded COP15 of the UN Framework Convention on Climate Change, would fail.

It is difficult, in retrospect, to convey the physical and emotional intensity of the experience. Following months of dire warnings about how this was the last chance to save humanity from runaway climate change, here we all were – on the brink. None of us had slept properly for at least a week. Every night negotiations dragged on into the small hours, and most national representatives were now visibly exhausted. A few scientists – vastly outnumbered by delegates, journalists and lobbyists – pottered around, talking to anyone who would listen and offering graphs of likely future temperature change resulting from the outcome of the negotiations. Their calculations suggested that the world bequeathed by the national pledges currently on the table at Copenhagen would warm by four degrees, perhaps more. We all knew what this meant. It meant planetary-scale destruction and perhaps a mortal threat to civilisation.

Much of the hot-house atmosphere at the negotiations arose because everyone knew how desperately important they were. Copenhagen was more than just another attempt at furthering global environmental governance. Here was humanity, meeting together in all our cultural and political diversity, trying to decide how to keep the planet's temperature within tolerable bounds. It was an awesome prospect; the human species pooling its collective intelligence in an effort to protect its only home. The only problem was, it wasn't working. We couldn't agree on anything. No one was prepared to give an inch in their well-rehearsed positions.

By a strange quirk of fortune, I was one of only fifty or so people who witnessed first-hand what really happened at the climax of Copenhagen – because I was in the room where the all-important final-hours heads of state negotiations were being conducted. I was not there as a journalist or campaigner: all of these were banished behind crowd-control barriers at the end of the corridor. Instead I was present in my part-time capacity as climate-change adviser to the president of the Maldives, Mohamed Nasheed, who was one of only twenty national leaders – selected to represent different regions and interest groups – taking part in the final closed-door stage of the two-week negotiations.

The room was small, suffocatingly hot and very overcrowded. Presidents and prime ministers sat jammed shoulder-to-shoulder around tables arranged in a square, each with a small microphone. I stood wedged amongst other officials in the second row. Directly in front was President Nasheed, whilst to his right sat Felipe Calderón of Mexico, Jacob Zuma of South Africa, Gordon Brown, then UK prime minister, and President Obama. The Danish prime minister, Lars Lokke Rasmussen, was chairing, whilst the Secretary-General of the United Nations, Ban Ki-moon, sat to his right. Even with all these high-powered decision-makers in the room, things were not going well. The Danish prime minister Rasmussen looked hassled. He kept mopping his brow and much of the time seemed lost for words.

I remember noting early on how the Chinese representative was not, as might be expected at such a high-level meeting, China's premier Wen Jiabao. Indeed, it later took me hours to definitively identify him as Yu Qingtai, Special Representative on Climate Change Negotiations at China's Foreign Affairs ministry.[1] Mr Yu was clearly a seasoned negotiator. He had perfected the art of subtle obstinacy, raising his hand each time he wanted to make an objection with an air of aggrieved but dutiful regret. And well he might, for the job of China's representative at this late-hour heads of state meeting in Copenhagen was very simple. He had to say no.

At issue was a draft 'Copenhagen Accord', produced by the Danish prime minister in his capacity as chair. Even though time was running out, Rasmussen was still trying to please everyone, believing always that with enough gentle encouragement the different parties could eventually be brought to agree. The Danes' draft Accord was a long way from meeting the climate change planetary boundary, but following a fortnight of rancorous and increasingly bitter negotiations I felt it was probably the best we were going to get. The worst outcome of all would have been a total collapse in the multilateral climate change process, spelling doom for the UN Climate Change Convention and the ultimate failure of fifteen years of international effort. In the ensuing mess, carbon markets would collapse and the big planned investments in low-carbon energy would drain away. Some kind of compromise would have to be reached.

The draft text, which was written in the usual barely comprehensible legalese of international treaties, suggested that global emissions should peak in 2020 or thereabouts. By mid-century CO_2 emissions would be cut by half, and industrialised countries – the so-called Annex 1 nations of the original 1992 UN Climate Convention – would shoulder 80 per cent of that burden. Developing countries, in recognition of their smaller contribution to the atmosphere's increasingly weighty greenhouse gas blanket, would not be asked to take on specific cuts, but could volunteer their own 'nationally appropriate mitigation actions' if they wanted to. It seemed like a reasonable compromise under the circumstances. Rasmussen looked hopefully around the room for support.

Everyone seemed satisfied. Had we succeeded? Then there was a discreet cough. 'My country cannot accept these numbers,' said Yu Qingtai, holding up his hand. The date – any date – for global peaking of emissions would put too much responsibility on developing countries to curb their growth, he insisted. Rich countries had caused the climate-change problem, and they must solve it without harming the legitimate development prospects of the poor. To get Chinese agreement, the 2020 year was excised, and replaced by woolly language stating that emissions must peak 'as soon as possible'. The Copenhagen Accord had been fatally weakened. But there was more to come. The discussion moved on to the mid-century 80 per cent target for industrialised-country cuts. Once again, with the same look of weary resignation, Mr Yu's hand went up. 'This is not acceptable,' he insisted. He was backed by Jairam Ramesh, the Indian environment minister, and also by Saudi Arabia – eager, as always, to derail the proceedings in the interests of selling more oil.

Angela Merkel, the German chancellor, was aghast. 'Why can't we mention even our own targets?' she demanded, looking pale and tired. Kevin Rudd, then prime minister of Australia, objected in even stronger terms, banging his microphone in frustration. France's Nicolas Sarkozy and Britain's Gordon Brown both made clear that these targets should stay. This was a 'red line' for them all, as the 80 per cent target had already been agreed at an earlier G8 meeting.[2] But it needed to be restated properly in a climate-change agreement if it was ever going to happen. Even Brazil, formally an ally of China in the negotiations, pointed out that the Chinese position was illogical – you can remove your own targets, but not someone else's. Speaking for the Maldives, President Nasheed reminded China that removing this all-important number would spell doom for low-lying and vulnerable island nations like his. With the prime minister of Grenada, on behalf of the Association of Small Island States, he insisted that the text should also retain a target for keeping global temperature rise below 1.5 degrees Celsius – against yet another Chinese objection.

Playing for time, Mr Yu asked for a break to consult his 'superiors' – begging the inevitable question of why China had sent a relatively junior official to negotiate in person with the president of the United States and other world leaders. Indeed, the implied snub was not lost on President Obama, who complained that it would have been better if he had been able to hold talks with someone actually empowered to take decisions. But there was nothing anyone could do, and the meeting broke up – at close to midnight, on the final evening of the conference – with the draft text still not agreed. I headed outside for some fresh air with President Nasheed, pursued by camera crews desperate for news. We said nothing, for there was nothing to say.

When the leaders reassembled, it was clear at once that the game was over. Neither the 80 per cent cut for rich countries, nor the mid-century 50 per cent cut, survived the night. The final Copenhagen Accord, stitched together as a face-saving measure by 2 am that morning, might still have been marginally better than nothing – but scientific analysts quickly warned that it would be unlikely to hold the global temperature rise below 3 or even 4 degrees higher than pre-industrial levels. When it was presented to the entire conference for adoption, chaos ensued. Some delegations whose leaders had not been invited to the heads of state meeting furiously denounced the draft accord as having been stitched up without their knowledge or approval. The majority, including the Maldives, were prepared to accept it as just one more small step along a much longer road, but such pragmatism failed to win over the angriest and loudest voices of opposition. It was midday on Saturday, twelve hours after the conference had been scheduled to close, before it was decided that the Copenhagen Accord would only be 'taken note of' by the assembled nations – and would therefore have no legal force within the UN climate process. It was the worst kind of compromise: one that satisfied nobody. People were not just disappointed, they were furious. And rightly so.

Copenhagen failed for many reasons. Principally, there were bitter divisions between rich and poor nations, each of whom blamed the other for failing to take leadership. Boxed in by its own domestic political paralysis, the United States was unable to make the concessions other states needed to see to prove its good faith. China, and also India, were flexing their muscles for the first time – showing beyond doubt that global agreements could not be made any longer without their support and cooperation. China was the big story: here was a new, emerging global superpower, going eyeball to eyeball with the United States – and winning. In the new reality of shifting geopolitics, China would be the key arbiter between success and failure. Decisions made in Beijing would count for as much, if not more, than decisions made in Washington. But with the ruling Communist Party dependent on double-digit growth rates for stability and control, China was not prepared to make commitments at the international level that would threaten its ability to fuel its economic miracle with coal.

Copenhagen showed us all what failing to meet a planetary boundary looks and feels like. But this failure need not be terminal. As I will show later, at a subsequent UN climate-change meeting at the end of 2010 in Cancun, a very different global politics was in evidence. Following the gloom of Denmark, today I believe that a way can be mapped out towards a deal on climate that will not only lead us more quickly towards the planetary boundary but will do so in a way that brings prosperity and clean growth to billions of people around the world. And perhaps most exciting of all is the fact that the country that is showing the way most forcefully and determinedly towards this better future is the same emerging superpower that blocked progress in Copenhagen: the People's Republic of China.

BACK TO THE BOUNDARIES

Based on the pioneering work of the 29 scientists making up the planetary boundaries expert group, this book has made the case that the Earth system has inherent ecological limits within it, and that seven out of nine of these limits can now be identified and quantified by science. The concept of a limited planet placing constraints on humanity is central to environmentalism, and we owe the Green movement a debt for making this philosophical case so strongly and ultimately persuasively – before science was able to confirm that ideas about ecological limits were well-founded. But as I have shown repeatedly, I differ from most Green thinkers in believing that in the short to medium term ecological limits need constrain neither our numbers as a species nor the growth of our economic activity. As the expert group wrote in its original 2009 *Nature* paper, 'as long as the thresholds are not crossed, humanity has the freedom to pursue long-term social and economic development'.[3] Our global civilisation can continue to flourish indefinitely within the 'safe operating space' provided by the planetary boundaries.

For easy reference, here are the planetary boundaries in summary, presented in the same order as chapters in this book and with those already crossed highlighted.

Earth system process	Control variable	Boundary	Pre-industrial	Latest data
Chapter 2: Biodiversity loss	Extinction rate, number of species per million per year	10	1	>100
Chapter 3: Climate change	Atmospheric CO_2 concentration, parts per million	350	280	391[4]
Chapter 4: Nitrogen cycle	Amount of N_2 removed from the atmosphere, millions of tonnes per year	35	0	121
Chapter 5: Land system change	Percentage of global land-cover converted to cropland, (millions of hectares)	15 (1,995)	5 (665)	11.7 (1,554)
Chapter 6: Global freshwater use	Consumptive use of withdrawn runoff, cubic kilometres per year	4,000	415	2,600
Chapter 7: Chemical pollution *Not yet quantified*	–	–	–	–
Chapter 8: Atmospheric aerosol loading *Not yet quantified*	–	–	–	–

Chapter 9: Ocean acidification	Global oceanic aragonite saturation ratio	2.75	3.44	2.90
Chapter 10: Stratospheric ozone depletion	Stratospheric O_3 concentration, Dobson units	276	290	283

Source: Data from J. Rockström et al, 2009: 'Planetary Boundaries: Exploring the Safe Operating Space for Humanity', *Ecology and Society*, 14,2, 32, Appendix 1, table S2, except where stated.

Several responses to the planetary boundaries are common enough to bear a quick rehearsal here. The first is: What about population? Why not set a boundary for a permissible number of humans? Firstly, this objection is moot because human population is a driver of environmental impact, not a qualifying physical Earth-system process in and of itself. But the objectors are quite correct in that a burgeoning population is likely to cause more environmental impact than a declining one. A 2010 scientific paper indeed makes clear that lower population scenarios by mid-century lead to significantly reduced carbon emissions as compared with alternative, higher-population outcomes.[5] But it does not necessarily imply, as the authors of that study are careful to point out, that promoting birth control should be a policy choice. That would be bad science as well as bad politics. For a start, it only follows that population reductions lead to emissions reductions if the factors affecting birth and death rates are themselves independent of economics. In the real world, the best way to reduce the growth in human populations is to encourage faster economic development, accelerated urbanisation, and therefore an earlier demographic transition to the lower birth rates already experienced in the most affluent societies. But faster economic growth will mean higher use of energy and more emissions, everything else remaining equal.

One way round this is to accept that population control must be authoritarian, as the environmentalist Jonathon Porritt appears to do in writing approvingly about China's one-child policy, which he calls 'the biggest single CO_2 abatement achievement since Kyoto'.[6]

Population enthusiasts like Porritt also tend to ignore the fact that the biggest driver of increasing human numbers has been better life expectancies thanks to economic progress and modern medical science. Should we therefore seek to restore death rates to their higher levels of yesteryear? Of course not. And of course, as well, we can all agree that access to family planning and female education – which tend to reduce birth rates in developing countries in advance of the demographic transition – are desirable in and of themselves. But they are desirable because they increase people's choices, not limit them. No one should be forced to have an unwanted child. But neither should anyone be forced not to have a wanted child – as has happened in China, where an intrusive state has intervened in the family-planning decisions of a billion people. The same goes for more gentle moral pressure too. I would never ask individuals, as Porritt does, to 'do your bit for addressing climate change by having fewer children – or even no children'.[7] To do so is an objectionable political intrusion into a highly individual private sphere. To be judged by Greens like Porritt on the basis of how many children one has is as personally insulting as it is counterproductive for the environmental movement.

As well as population, another frequently heard objection I have come up against is that the planetary boundaries do not deal with resource constraints, which were central to earlier thinking about ecological limits like the Club of Rome's groundbreaking 1972 report *Limits to Growth*. Again, this is to misunderstand the physical and ecological nature of the proposed boundaries: it makes no difference to the biosphere if humans run out of iron, for example. Nor does it make any difference if we use up all the cheaply extractable oil – as has recently become the concern of the 'peak oil' crowd – except to the extent that humanity's response to declining oil supplies, like burning more coal or extracting more tar sands, will negatively affect real planetary boundaries like climate change. But peak oil might also be a good thing if it adds to rising prices of fossil fuels sufficiently to encourage the faster uptake of low- and zero-carbon alternatives. Either way, the planetary boundaries are the metric by which humanity's response to resource crunches should be judged – they are not concerned with resource shortages in and of themselves.

Third, to the biggest and most central concern of all: economic growth. Throughout this book I have referred approvingly to growth, technology and innovation as ways to solve pressing environmental challenges. But won't growth go on to cause even more problems than it solves? And is ever-increasing consumption even possible on a physically limited planet? De-growth, on the other hand, can be environmentally beneficial: industrialised countries saw declines in carbon emissions following the economic recession that began in 2008. These are strong arguments, and need to be looked at carefully. But I believe my pro-growth perspective holds up to the challenge, for several equally strong reasons.

First, growth in consumption, like growth in population, tends to top out once a certain level is achieved. After all, there is only so much that any one person can eat – and once your fridge is already groaning with fresh and tempting produce from around the globe, a second and third equally well-stocked fridge is not particularly desirable. In economics language, there is a declining 'marginal utility' applying to further consumption. The same trend can be seen in any area: in China, vehicle ownership is exploding – but in the more prosperous US, a saturated car market cannot grow much more now that households on average already have two cars. Moreover, once a certain level of prosperity is reached, people tend to shift their concerns away from consumption and towards other areas of life satisfaction. It is no accident that environmental groups like the New Economics Foundation, which worry about the psychological and social evils of overconsumption, only flourish in rich countries. To a semi-destitute family picking over a rubbish dump on the outskirts of Manila, such concerns must seem as irrelevant as they are self-indulgent.

This leads on to my second point: that there can and should be no argument against rapid and sustained growth throughout the developing world. By definition, reducing poverty means raising levels of consumption – and not just of basic necessities like food and clean water, but of 'luxury' goods like mobile phones and travel too. In speaking to British audiences I have often come across the rather quaint notion that people in poorer countries only want to get more prosperous so they can 'be like us', having been exposed even in their shacks, presumably, to the insidious consumerist propaganda of satellite television. The implication is that if 'we' could consume less, then 'they' would again emulate us by limiting their aspirations for development and prosperity, thereby helping the environment. This is a particular favourite of well-meaning, environmentally aware middle-class audiences in prosperous areas.

Of course, in reality the idea is as fanciful as it is patronising. The truth is that people in poorer countries – to the extent that one can generalise – are desperate to increase their security by having access to modern medicine, fair employment, capital, and a whole host of other things that those of us living in rich countries have taken for granted since birth. Since abandoning the failed experiment of Maoism, China's double-digit rates of economic growth have lifted hundreds of millions of people out of poverty in what is surely the greatest development success story of modern times. Over the last decade, having abandoned its own version of state socialism, India has begun to follow suit – and both countries are increasingly acknowledged as emerging economic and geopolitical superpowers. In contrast, sub-Saharan Africa has been somewhat left behind, though there too countries with freer economies like Rwanda and Ghana are beginning to experience welcome growth spurts.

China is already an environmental superpower, of course. As Oxford University's Dieter Helm puts it: 'Whether China builds 1,000 gigawatts of coal-fired electricity generation and whether it adds half a billion cars with conventional engines is of an order of magnitude more important to climate change than virtually any other trend.'[8] China's heavy use of coal makes it a prodigious emitter: even on a per capita basis, China is rapidly catching up with the France and Sweden (which have some of the lowest per capita emissions in Western Europe thanks to nuclear and hydroelectric power) and may already have overtaken them by the time you read this. Middle-class British audiences are quite right to worry about China's rise, for their own well-intentioned efforts to reduce their household emissions by turning down thermostats and walking their kids to school shrink quickly into irrelevance by comparison. Just consider that in 2009, whilst emissions fell in the US and Europe thanks to the recession, China added more to its CO_2 output in a single year than the entire emissions of the UK, Spain and Ireland combined (about 900 million tonnes).[9] Opposing growth in developing countries is a non-starter, however. All we can do is try to assist them in ensuring that their rising consumption of energy can come as much as possible from low-carbon sources, and that other planetary boundaries are considered in sustainable development plans.

My third concern about attacks on the concept of economic growth is that there really does seem to be no conceivable alternative at present. Environmental economists like Herman Daly have struggled for decades to design a viable macroeconomic system that does not require growth as a central principle – and have palpably failed, as the economist Tim Jackson candidly admits in his thoughtful 2009 treatise on the subject, *Prosperity Without Growth*. The alternative to growth in the modern market system is painful contraction, unemployment and political instability, as numerous recessions since the 1930s have demonstrated. Given the need for interest to be paid on capital, and for increases in labour productivity to be balanced out by higher levels of overall production, growth is central to successful capitalist economies. No one argues, however, that the kind of rapid double-digit rates of growth seen today in successful emerging economies are appropriate for the likes of Europe, Japan or North America. In developed countries, 2–3 per cent per year is plenty. This differential is helping to reduce global inequality too, in stark contrast to the fears expressed a few years ago by the anti-globalisation movement. But even in rich countries, zero growth is not a viable option.

Another piece of good news is that as economies grow they tend to become less resource-intensive per unit of output. In other words, we are constantly getting relatively more efficient in our use of the world's resources even as the overall level of human consumption grows. This trend towards de-materialisation is positive for several planetary boundaries. In the area of nitrogen, for example, Chinese food production rose by nearly 200 per cent between 1981 and 2007, for only a 50 per cent increase in fertiliser.[10] Another study, looking at the same multi-decadal period, found that a 45 per cent more affluent world used only 22 per cent more crops and 13 per cent more energy.[11]

Of course, in both these cases, absolute resource use went up even as relative use went down – because of economic growth. But this is not always the case. Some basic resources are even being used at lower absolute levels as humanity gets more affluent: between 1980 and 2006, for instance, a richer world actually used 20 per cent less wood.

Looking further out into the future, it is perhaps possible to envisage a world economy that enjoys constant growth even as its use of materials is static or even declining, thanks to de-materialisation. Technology will help: in consuming music electronically via downloads rather than plastic CDs, we use less oil. E-books and online information dissemination will hopefully eventually reduce paper consumption too. At a conceptual level, what we must surely aim for is a closed-loop economy, where rates of recycling come as close to 100 per cent as practically possible, and what is not recycled can be regenerated naturally within the biosphere. (This recycling may vary from low-tech, via people sorting their rubbish, to high-tech, using plasma-arc reactors.) Within such a system, consumption rates can still rise as materials circulate faster. The only external ingredient that must also increase in order to drive this is energy: and energy is not limited at all in any fundamental thermodynamic sense, if nuclear fission (on Earth) and fusion (in the sun) are utilised sustainably. The planetary boundaries provide a physical and ecological limit to how far humans can trespass on the biosphere, but if we respect them fully then I believe that growth – as currently conceived – can continue more or less indefinitely. In contrast to many environmentalists, I do not see any convincing ecological reason why everyone in the world should not be able to enjoy rich-country levels of prosperity over the half-century to come. None of the planetary boundaries rule out this leap forward in human development – and as far as the Earth system is concerned, it is the boundaries that must provide the ultimate guidelines to the human project.

Humanity has so far transgressed only three of the nine planetary boundaries – biodiversity, climate and nitrogen – and has successfully navigated away from a breach of one, the ozone layer. Boundaries on ocean acidification, land use and water use are still avoidable. I am confident that we can respect them, and move back into the safety zone with the first three, without the need to try and limit human aspirations. I hope this book has shown convincingly that whilst ecological limits are real, economic limits are not. We can respect all the boundaries at the same time as eliminating poverty, allowing for a peak in world population around mid-century, and within the context of an economic system that requires growth.

The transition of humanity towards a sustainable presence in the Earth system will constitute an epochal event, equivalent at least to the Industrial Revolution that so transformed our civilisation over the last two centuries. It represents an enormous opportunity for business, as happened with ozone regulation, and can increasingly be seen in the sphere of climate change. From being concerned that cutting carbon emissions will damage competitiveness, many governments are beginning to see going low-carbon as the very engine of future growth and competitiveness that they are looking for. Led by the Maldives, a growing number of developing countries are pledging themselves to carbon neutrality, now including Costa Rica, Ethiopia and Samoa, and many others have upped their levels of ambition in going for low-carbon status. Many small island states and least-developed countries, fed up with being seen as silent victims of climate change, are determined to become the low-carbon winners of a more innovative future. This change in narrative has begun to transform the climate-change negotiations: in Cancun, amidst an atmosphere of unprecedented trust and good will, 193 countries agreed to put the process back on track – laying to rest the ghost of Copenhagen. Central to this success has been the quiet emergence of a new alliance, the Cartagena Dialogue, which brings together upwards of 30 countries – rich and poor, large and small – in a shared ambition to increase the pace of global action on climate change.

And here is the really big story. China's emissions may be rising, but its investment in low-carbon technologies leads the world. In September 2010 China overtook the stalling US as a destination for renewable investment. 'China has all the benefits of capital, government will, and it's a massive market,' gushed an analyst from Ernst & Young.[12] China has already overtaken the US in terms of installed wind-energy capacity, and in 2010 added a colossal 16 gigawatts

– nearly half the entire world total of new wind installations.[13] Despite the shock of the Fukushima accident, I would expect China to quickly resume its heavy investment in new nuclear generating capacity, which is likely to include launching a research programme on cleaner fourth-generation nuclear power using thorium reactors. Many in the US are increasingly rattled about being overtaken in the clean-energy race: not for nothing did President Obama's 2011 State of the Union address raise the spectre of a 'Sputnik moment', with China now in the place of the old Soviet Union.

The US is being left behind, and for one reason: it refuses to be bound by clean-energy regulation. Without the political driver of new markets, technological innovation in low-carbon alternatives cannot take off. In contrast, China has aggressive targets both for the reduction in carbon emissions per unit of GDP and for the penetration of renewables and nuclear into the grid. China may have played a wrecking game in Copenhagen, but since then the country has proved that it is deadly serious about dealing with climate change – and winning massive economic gains in the process. As I argued earlier, strong and binding international agreements will be vital if this clean-energy revolution is to continue to accelerate globally – and this must include the emerging big emitters just as it does the rich countries, perhaps in a revamped and expanded version of the Kyoto Protocol.

Equally strong action will be needed in other areas if equivalent progress is to be made in meeting and respecting the other planetary boundaries too. Whilst 2010 saw some hope that biodiversity is beginning to become a global policy concern on the scale of climate change, science and policy on the nitrogen cycle is in its infancy, and water and land use are hardly considered at all on a global scale. And interlinkages between all the boundaries must be recognised and addressed through any global multilateral process: already the Montreal and Kyoto Protocols are beginning to clash. As we have seen, there are many potential synergies too: encouraging denitrification means protecting wetland, thereby helping meet the water and biodiversity boundaries at the same time.

Can humanity manage the planet – and itself – towards this transition to sustainability? I believe that we can. Whether we will remains to be seen. But the grounds for optimism are at least as strong as the grounds for pessimism, and only optimism can give us the motivation and passion we will need to succeed. Voices of doom may be persuasive, but theirs is a counsel of despair. The world – and our own children – deserve better. The truth is that global environmental problems are soluble. Let us go forward and solve them.

NOTES

Introduction

1. http://www.grist.org/article/2010-06-30-gingrich-slams-obama-on-bp-gulfspill-and-sounds-off-on-climate/
2. I swear I wrote the bit about the volcano months before I discovered Gingrich's quote.
3. BusinessGreen, 2011: 'Analysts: German nuclear shutdown set to send emissions soaring', 17 March 2011, http://www.businessgreen.com/bg/news/2034702/analysts-german-nuclear-shutdown-set-send-emissions-soaring
4. This is not an original insight of mine. I have heard it many times: James Lovelock may have said it first.

1: The Ascent of Man

1. Such carbonates were conclusively identified by the Mars Spirit Rover. See R. Morris et al., 2010: 'Identification of Carbonate-Rich Outcrops on Mars by the Spirit Rover', Science, 23 July 2010, vol. 329, no. 5990, pp. 421–4.
2. H. Williams and T. Lenton, 2007: 'The Flask model: emergence of nutrient-recycling microbial ecosystems and their disruption by environment-altering "rebel" organisms', Oikos, 116, 7, 1087–1105.
3. J. McNeil, 2001: Something New Under the Sun – An environmental history of the twentieth century world, Penguin, p.9
4. For the latest count, see the Population Clock on http://math.berkeley.edu/~galen/popclk.html
5. S. Brand, 2009: Whole Earth Discipline – an ecopragmatist manifesto, Atlantic Books.
6. R. Jastrow and M. Rampino, 2008: Origins of Life in the Universe, Cambridge University Press.

7. R. Cowen, 2004: History of Life, Blackwell Publishing, p. 290.
8. Jastrow and Rampino, cited above, p. 362.
9. F. Burton, 2009: Fire: The Spark That Ignited Human Evolution, University of New Mexico Press.

2: *The Biodiversity Boundary*

1. IUCN Red List 2008, Table 7 (excel spreadsheet): Species changing IUCN Red List status, from http://iucn.org/about/work/programmes/species/red_list/2008_red_list_summary_statistics/
2. S. Butchart et al., 2010: 'Global Biodiversity: Indicators of Recent Declines', Science, 328, 5982, 1164–8.
3. M. von Arx and C. Breitenmoser-Wursten, 2008: Lynx pardinus. In: IUCN 2009. IUCN Red List of Threatened Species. Version 2009.1. www.iucnredlist.org. Downloaded on 19 October 2009.
4. Secretariat of the Convention on Biological Diversity, 2010: Global Biodiversity Outlook 3, Montréal, 94 pp.
5. See http://assets.panda.org/downloads/wwf_tigers_e_1.pdf and IUCN listing for Panthera tigris for the latest information.
6. J. Walston et al., 2010: 'Bringing the Tiger Back from the Brink – The Six Percent Solution', PLoS Biology, 8, 9, e1000485.
7. K. Herrera et al., 2009: 'To what extent did Neanderthals and modern humans interact?', Biological Reviews, 84, 2, 245–57.
8. G. Stix, 2008: 'Traces of a Distant Past', Scientific American, July 2008, 56–63.
9. C. Finlayson et al., 2006: 'Late survival of Neanderthals at the southernmost extreme of Europe', Nature, 443, 850–3.
10. S. Churchill, 2009: 'Shanidar 3 Neandertal rib puncture wound and paleolithic weaponry', Journal of Human Evolution, 57, 2, 163–78.
11. F. Rozzi et al., 2009: 'Cutmarked human remains bearing Neandertal features and modern human remains associated with the Aurignacian at Les Rois', Journal of Anthropological Sciences, 87, 153–85.
12. R. McKie, 2009: 'How Neanderthals met a grisly fate: devoured by humans', The Guardian, 17 May 2009, http://www.guardian.co.uk/science/2009/may/17/neanderthals-cannibalism-anthropological-sciences-journal
13. R. Cowen, 2004: History of Life, Blackwell Publishing, p. 292.
14. C. Turney et al., 2008: 'Late-surviving megafauna in Tasmania, Australia, implicate human involvement in their extinction', PNAS, 105, 34, 12150–3.
15. J. Diamond, 2000: 'Blitzkrieg Against the Moas', Science, 287, 5461, 2170–1.
16. Cowen, cited above, p. 309.
17. C. Donlan, 2007: 'Restoring America's big wild animals', Scientific American, June 2007, 70–7.

18. C. Johnson, 2009: 'Ecological consequences of Late Quaternary extinctions of megafauna', Proceedings of the Royal Society B, 276, 1667, 2509–19.
19. C. Roberts, 2007: The Unnatural History of the Sea: The Past and Future of Humanity and Fishing, Gaia Books, p. 99.
20. Census of Marine Life, 2009: 'Describing Ocean Life in Olden Days, Researchers Upend Modern Notions of "Natura"' Animal Sizes, Abundance', Highlights News Release, http://www.coml.org/comlfiles/press/CoML_Oceans_Past_Public_Release_05.23.pdf
21. S. Reilly et al., 2008: Eubalaena glacialis. In: IUCN 2009. IUCN Red List of Threatened Species. Version 2009.1. www.iucnredlist.org. Downloaded on 21 September 2009.
22. J. McNeill, 2001: Something New Under the Sun: An Environmental History of the Twentieth-Century World, W. W. Norton & Company.
23. Roberts,cited above, p. 43.
24. See the Extinction Website (a fascinating resource): http://www.petermaas.nl/extinct/speciesinfo/greatauk.htm
25. Roberts, cited above, p. 74.
26. Roberts, p. 115.
27. Roberts, p. 69.
28. Roberts, p. 112.
29. B. MacKenzie et al., 2009: 'Impending collapse of bluefin tuna in the northeast Atlantic and Mediterranean', Conservation Letters, 2, 1, 26–35.
30. 'Illegal Italian spotter planes caught hunting down Mediterranean bluefin tuna', Wildlife Extra, June 2008, http://www.wildlifeextra.com/go/news/bluefin-tuna745.html#cr
31. M. Hickman, 2009: 'Revealed: the bid to corner world's bluefin tuna market', The Independent, 3 June 2009, http://www.independent.co.uk/environment/nature/revealed-the-bid-to-corner-worlds-bluefin-tuna-market-1695479.html
32. A. Aguirre-Munoz et al., 2008: 'High-impact Conservation: Invasive Mammal Eradications from the Islands of Western Mexico', Ambio, 37, 2, 101–7.
33. C. Hallman et al., 2008: 'Community dynamics of anaerobic bacteria in deep petroleum reservoirs', Nature Geoscience, 1, 569–70.
34. T. Reichenbach et al., 2007: 'Mobility promotes and jeopardizes biodiversity in rock–paper–scissors games', Nature, 448, 1046–9.
35. A. Chapman, 2009: Numbers of Living Species in Australia and the World, Report for the Australian Biological Resources Study, Canberra, Australia, September 2009, http://www.environment.gov.au/biodiversity/abrs/publications/other/species-numbers/2009/pubs/nlsaw-2nd-complete.pdf

36. R. Booth, 2009: 'Lost world of fanged frogs and giant rats discovered in Papua New Guinea', The Guardian, 7 September 2009, http://www.guardian.co.uk/environment/2009/sep/07/discovery-species-papua-new-guinea
37. E. Wilson, 2002: The Future of Life, Knopf, p. 124.
38. F. Ferretti et al., 2008: 'Loss of Large Predatory Sharks from the Mediterranean Sea', Conservation Biology, 22, 4, 952–64.
39. R. Myers et al., 2007: 'Cascading Effects of the Loss of Apex Predatory Sharks from a Coastal Ocean', Science, 315, 5820, 1846–50.
40. A. Bundy and P. Flanning, 2005: 'Can Atlantic cod (Gadus morhua) recover? Exploring trophic explanations for the non-recovery of the cod stock on the eastern Scotian Shelf, Canada', Canadian Journal of Fisheries and Aquatic Sciences, 62, 1474–89.
41. W. Ripple and R. Beschta, 2007: 'Restoring Yellowstone's aspen with wolves', Biological Conservation, 138, 3–4, 514–19.
42. H. Lessios, 1988: 'Mass mortality of Diadema Antillarum in the Caribbean: What have we learned?', Annual Review of Ecology and Systematics, 19, 371–93.
43. T. Hughes, 1994: 'Catastrophes, Phase Shifts, and Large-Scale Degradation of a Caribbean Coral Reef, Nature, 265, 1547–51.
44. T. Gardner et al., 2003: 'Long-Term Region-Wide Declines in Caribbean Corals', Science, 301, 5635, 958–60.
45. J. Diamond, 2001: 'Dammed Experiments!', Science, 294, 5548, 1847–8.
46. W. Turner et al., 2007: 'Global Conservation of Biodiversity and Ecosystem Services', BioScience, 57, 10, 868–73.
47. L. Burke and J. Maidens, 2004: Reefs at risk in the Caribbean, World Resources Institute.
48. B. Martin-Lopez et al., 2008: 'Economic Valuation of Biodiversity Conservation: the Meaning of Numbers', Conservation Biology, 22, 3, 505–809.
49. M. Toman, 1998: 'Why not to calculate the value of the world's ecosystem services and natural capital', Ecological Economics, 25, 1, 57–60.
50. R. Costanza et al., 1997: 'The value of the world's ecosystem services and natural capital', Nature, 387, 253–60.
51. http://bankofnaturalcapital.com/category/stocks-investments/
52. TEEB, 2010: The Economics of Ecosystems and Biodiversity: Mainstreaming the Economics of Nature: A synthesis of the approach, conclusions and recommendations of TEEB.
53. UNEP, 2010: 'TEEB Report Puts World's Natural Assets on the Global Political Radar', press release, 20 October 2010.
54. TEEB, 2010.
55. See http://maluabank.com/products.html

56. M. Crow and K. Kate, 2010: 'Biodiversity offsets: policy options for government', BBOP Secretariat (draft publication).
57. World Bank and FAO, 2009: 'The sunken billions: The economic justification for fisheries reform', Agriculture and Rural Development Department, The World Bank, Washington D.C.
58. UN Convention on Biodiversity, 2010: Decision X/2 – The Strategic Plan for Biodiversity 2011–2020 and the Aichi Biodiversity Targets.

3: The Climate Change Boundary

1. J. Harries et al., 2001: 'Increases in greenhouse forcing inferred from the outgoing longwave radiation spectra of the Earth in 1970 and 1997', Nature, 410, 355–7.
2. NASA, 'NASA Research Finds 2010 Tied for Warmest Year on Record', 12 January 2011, http://www.nasa.gov/topics/earth/features/2010-warmest-year.html
3. See 2009 State of the Climate highlights brochure on http://www1.ncdc.noaa.gov/pub/data/cmb/bams-sotc/2009/bams-sotc-2009-brochure-lo-rez.pdf
4. N. Gillett et al., 2003: 'Detection of human influence on sea level pressure', Nature, 422, 292–4.
5. V. Ramaswamy et al., 2006: 'Anthropogenic and Natural Influences in the Evolution of Lower Stratospheric Cooling', Science, 311, 5764, 1138–41.
6. B. Santer et al., 2003: 'Contributions of Anthropogenic and Natural Forcing to Recent Tropopause Height Changes', Science, 301, 5632, 479–83.
7. D. Seidel et al., 2008: 'Widening of the tropical belt in a changing climate', Nature Geoscience, 1, 21–4.
8. See X. Zhang et al., 2007: 'Detection of human influence on twentieth-century precipitation trends', Nature, 448, 461–5, and R. Allan and B. Soden, 2008: 'Atmospheric Warming and the Amplification of Precipitation Extremes', Science, 321, 5895, 1481–4.
9. A. Sibley, 2010: 'Analysis of extreme rainfall and flooding in Cumbria 18–20 November 2009', Weather, 65, 287–92.
10. R. Boswell, 2010: 'Another big-ice Arctic thaw, say experts', The Vancouver Sun, 8 September 2010, http://www.vancouversun.com/technology/Another+Arctic+thaw+experts/3496268/story.html
11. R. Kwok et al., 2009: 'Thinning and volume loss of the Arctic Ocean sea ice cover: 2003–2008', Journal of Geophysical Research, 114, C07005.
12. NSIDC, 'Arctic Oscillation brings record low January extent, unusual mid-latitude weather', 2 February 2011, http://nsidc.org/arcticseaicenews/2011/020211.html

13. M. Wang and J. Overland, 2009: 'A sea ice free summer Arctic within 30 years?', Geophysical Research Letters, 36, L07502.
14. T. Cronin et al., 2010: 'Quaternary Sea-ice history in the Arctic Ocean based on a new Ostracode sea-ice proxy', Quaternary Science Reviews, in press.
15. V. Petoukhov and V. A. Semenov, 2010: 'A link between reduced Barents-Kara sea ice and cold winter extremes over northern continents', J. Geophys. Res., 115, D21111.
16. M.A. Moline et al., 2008: 'High Latitude Changes in Ice Dynamics and Their Impact on Polar Marine Ecosystems', Annals of the New York Academy of Sciences, 1134, 267–319.
17. G. Durner et al., 2009: 'Predicting 21st-century polar bear habitat distribution from global climate models', Ecological Monographs, 79, 25–58.
18. See http://www.doi.gov/news/08_News_Releases/080514a.html. The subtitle to the press release is 'Rule will allow continuation of vital energy production in Alaska'.
19. J. Hansen et al., 2008: 'Target Atmospheric CO_2: Where Should Humanity Aim?', Open Atmospheric Science Journal, 2, 217–31.
20. J. Overpeck and B. Udall, 2010: 'Dry Times Ahead', Science, 328, 5986, 1642–3.
21. See 2007 IPCC Fourth Assessment Report, WGII, p. 622.
22. C. Allen et al., 2010: 'A global overview of drought and heat-induced tree mortality reveals emerging climate change risks for forests', Forest Ecology and Management, 259, 4, 660–84.
23. H. Adams et al., 2010: 'Climate-Induced Tree Mortality: Earth System Consequences', Eos, Transactions, 91, 17.
24. D. Biello, 2008: 'Global warming's first mammal victim?', News Blog, Scientific American, 4 December 2008, http://www.scientificamerican.com/blog/60-second-science/post.cfm?id=global-warmings-first-mammal-victim-2008-12-04
25. C. Raxworthy et al., 2008: 'Extinction vulnerability of tropical montane endemism from warming and upslope displacement: a preliminary appraisal for the highest massif in Madagascar', Global Change Biology, 14, 8, 1703–20.
26. J.-M. Robine et al., 2008: 'Death toll exceeded 70,000 in Europe during the summer of 2003', C. R. Biologies 331.
27. G. Jones et al., 2008: 'Human contribution to rapidly increasing frequency of very warm Northern Hemisphere summers', Journal of Geophysical Research, 113, D02109.
28. 'Japan endures hottest summer on record', Washington Post, 2 September 2010.

29. F. Kuglitsch et al., 2010: 'Heat wave changes in the eastern Mediterranean since 1960', Geophysical Research Letters, 37, L04802.
30. K. Petrone et al., 2010: 'Streamflow decline in southwestern Australia, 1950–2008', Geophysical Research Letters, 37, L11401.
31. T. Lenton et al., 2008: 'Tipping elements in the Earth's climate system', PNAS, 105, 6, 1786–93.
32. I. Velicogna, 2009: 'Increasing rates of ice mass loss from the Greenland and Antarctic ice sheets revealed by GRACE', Geophysical Research Letters, 36, L19503.
33. J. Chen et al., 2009: 'Accelerated Antarctic ice loss from satellite gravity measurements', Nature Geoscience, 2, 859–62.
34. A. Cazenave and W. Llovel, 2010: 'Contemporary Sea Level Rise', Annual Review of Marine Science, 2, 145–73.
35. S. Jevrejeva et al., 2010: 'How will sea level respond to changes in natural and anthropogenic forcings by 2100?', Geophysical Research Letters, 37, L07703.
36. S. Rahmstorf et al., 2007: 'Recent Climate Observations Compared to Projections', Science, 316, 5825, 709.
37. T. Kuhlbrodt et al., 2009: 'An Integrated Assessment of changes in the thermohaline circulation', Climatic Change, 96, 489–537.
38. Y. Malhi et al., 2009: 'Exploring the likelihood and mechanism of a climate-change-induced dieback of the Amazon rainforest', PNAS, 106, 49, 20610–15.
39. E. Schuur et al., 2008: 'Vulnerability of Permafrost Carbon to Climate Change: Implications for the Global Carbon Cycle', Bioscience, 58, 8.
40. C. Tarnocai et al., 2009: 'Soil organic carbon pools in the northern circumpolar permafrost region', Global Biogeochemical Cycles, 23, GB2023.
41. A. Bloom et al., 2010: 'Large-Scale Controls of Methanogenesis Inferred from Methane and Gravity Spaceborne Data', Science, 327, 5963, 322–5.
42. N. Shakhova et al., 2010: 'Extensive Methane Venting to the Atmosphere from Sediments of the East Siberian Arctic Shelf', Science, 327, 5970, 1246–50, and G. Westbrook et al., 2009: 'Escape of methane gas from the seabed along the West Spitsbergen continental margin', Geophysical Research Letters, 36, L15608.
43. http://www.realclimate.org/index.php/archives/2010/03/arctic-methane-on-the-move/
44. D. Archer et al., 2009: 'Ocean methane hydrates as a slow tipping point in the global carbon cycle', PNAS, 106, 49, 20596–601.
45. E. Dlugokencky et al., 2009: 'Observational constraints on recent increases in the atmospheric CH4 burden', Geophysical Research Letters, 36, L18803.

46. J. Steffensen et al., 2008: 'High-Resolution Greenland Ice Core Data Show Abrupt Climate Change Happens in Few Years', Science, 321, 5889.
47. A. Brauer et al., 2008: 'An abrupt wind shift in western Europe at the onset of the Younger Dryas cold period', Nature Geoscience, 1, 8.
48. E. Rohling et al.: 2009: 'Antarctic temperature and global sea level closely coupled over the past five glacial cycles', Nature Geoscience, 2, 500–4.
49. R. DeConto et al., 2008: 'Thresholds for Cenozoic bipolar glaciation', Nature, 455, 652–6.
50. A. Tripati et al., 2009: 'Coupling of CO_2 and Ice Sheet Stability Over Major Climate Transitions of the Last 20 Million Years', Science, 326, 5958, 1394–7.
51. T. Naish et al., 2009: 'Obliquity-paced Pliocene West Antarctic ice sheet oscillations', Nature, 458, 322–8.
52. J. Hansen et al., 2008: 'Target atmospheric CO_2: Where should humanity aim?', Open Atmospheric Science Journal, 2, 217–31.
53. P. Friedlingstein et al., 2010: 'Update on CO_2 emissions', Nature Geoscience, 21 November 2010.
54. R. Pielke Jnr, 2010: The Climate Fix, Basic Books, pp. 46 and 49.
55. http://www.neweconomics.org/sites/neweconomics.org/files/Well-being_and_the_Environment.pdf
56. http://www.neweconomics.org/sites/neweconomics.org/files/The_Happy_Planet_Index_2.0_1.pdf
57. Pielke, cited above, Table 4.4, p. 114.
58. World Nuclear Association, 'China Nuclear Power', http://www.world-nuclear.org/info/inf63.html
59. Each 500-MW coal plant produces 3 million tonnes of CO_2. Massachusetts Institute of Technology study, 'The Future of Coal', http://web.mit.edu/coal/The_Future_of_Coal_Summary_Report.pdf
60. S. Postel and B. Richter, 2003: Rivers for Life: Managing Water For People And Nature, Island Press.
61. Nuclear Power in France, World Nuclear Association, http://www.world-nuclear.org/info/inf40.html
62. Royal Academy of Engineering, 'The Cost of Generating Electricity', http://www.raeng.org.uk/news/publications/list/reports/Cost_Generation_Commentary.pdf
63. 'Nuclear construction builds up', World Nuclear News, 4 January 2011.
64. http://www.desertec.org/en/concept/questions-answers/
65. http://www.energy.ca.gov/siting/solar/index.html
66. http://www.desertec-australia.org/content-oz/australiacsppotential.html
67. Bloomberg, 'Australia Overtakes U.S. in Per Capita CO_2 Emissions', 9 September 2009.

68. Reuters, 'Australian Carbon Emissions Set To Rise: Government', 9 February 2011.
69. GWEC, 'Global wind capacity to reach close to 200 GW this year', 23 September 2010.
70. EIA, Table H4, 'World installed coal-fired generating capacity by region and country, 2007–2035', accessible via http://www.eia.doe.gov/oiaf/ieo/ieoecg.html
71. H. Trabish, 2010: 'Wind's 2010 Top Ten: Bust and Building, Coming and Going', greentechmedia.com, 29 December 2010.
72. H. Trabish, 2010: 'What the Big Atlantic Backbone Transmission Buy-In Means for U.S. Offshore Wind', greentechmedia.com, 25 October 2010.
73. http://www.businessgreen.com/bg/news/1870086/fierce-opposition-drives-wind-farm-approvals-low
74. Public Interest Research Centre on behalf of The Offshore Valuation Group, 2010: The Offshore Valuation: A valuation of the UK's offshore renewable energy resource.
75. http://www.businessgreen.com/bg/news/1805813/uk-cuts-ribbon-worlds-largest-offshore-wind-farm
76. The Economist, 'Power to the European market', 11 November 2010.
77. http://www.worldfuturecouncil.org/fileadmin/user_upload/Rob/press/publications/WFC_Academic_Paper_New_Money.pdf
78. IEA, 'Global gaps in clean energy RD&D', 2010, http://www.iea.org/papers/2010/global_gaps.pdf
79. I. Galiana and C. Green, 2009: 'Let the global technology race begin', Nature, 462, 570–1.
80. http://www.sandbag.org.uk/site_media/pdfs/reports/caportrap.pdf
81. Reuters, 'As hybrid cars gobble rare metals, shortage looms', 31 August 2009.
82. Argonne National Laboratory, 'An introduction to ANL's IFR program', http://web.archive.org/web/20071009064447/www.nuc.berkeley.edu/designs/ifr/anlw.html
83. A 1-GW IFR would use about a tonne of DU per year. There are 686,500 tonnes sitting in storage barrels in Portsmouth, Ohio, and Paducah, Kentucky. The US has about 1000 GW of installed electrical generation capacity.
84. http://declaration.klimaforum.org/files/declaration/declaration_screen.pdf
85. See http://www.us-cap.org/
86. See http://trillionthtonne.org
87. M. Allen et al., 2009: 'The exit strategy', Nature Reports Climate Change, 30 April 2009.

4: The Nitrogen Boundary

1. Encyclopaedia Britannica, entry for 'famine': http://www.britannica.com/EBchecked/topic/201392/famine
2. J. Galloway, 2003: 'The Nitrogen cascade', BioScience, 53, 4, 341–56.
3. A. Maddison, HS–8: The World Economy, 1–2001 AD, http://www.ggdc.net/maddison/other_books/HS-8_2003.pdf
4. W. Crookes, 1899: The Wheat Problem, G. P. Putnam's Sons.
5. V. Smil, 2004: Enriching the Earth: Fritz Haber, Carl Bosch, and the Transformation of World Food Production, The MIT Press, introduction.
6. See full lecture on http://nobelprize.org/nobel_prizes/chemistry/laureates/1918/haber-lecture.pdf
7. Smil, cited above, p. 103.
8. Smil, p.105.
9. M. Goran, 1967: The story of Fritz Haber, University of Oklahoma Press.
10. J. Erisman et al., 2008: 'How a century of ammonia synthesis changed the world', Nature Geoscience, 1, 636–9.
11. J. Jackson, 2008: 'Ecological extinction and evolution in the brave new ocean', PNAS, 105, Supplement 1, 11458–65.
12. R. Diaz and R. Rosenberg, 2008: 'Spreading Dead Zones and Consequences for Marine Ecosystems', Science, 321, 5891, 926–9.
13. F. Magnani et al., 2008: 'Magnani et al. reply', Nature, 451, E3–E4.
14. C. Goodale et al., 2002: 'Forest carbon sinks in the Northern Hemisphere', Ecological Applications, 12, 3, 891–9.
15. C. Clark and D. Tilman, 2008: 'Loss of plant species after chronic low-level nitrogen deposition to prairie grasslands', Nature, 451, 712–15.
16. R. Duce et al., 2008: 'Impacts of Atmospheric Anthropogenic Nitrogen on the Open Ocean', Science, 320, 5878, 893–7.
17. A. Oczkowski et al., 2009: 'Anthropogenic enhancement of Egypt's Mediterranean fishery', PNAS, 106, 5, 1364–7.
18. Clark and Tilman, cited above.
19. C. Stevens et al., 2004: 'Impact of Nitrogen Deposition on the Species Richness of Grasslands', Science, 303, 5665, 1876–9.
20. Clark and Tilman, cited above.
21. E. Davidson, 2009: 'The contribution of manure and fertilizer nitrogen to atmospheric nitrous oxide since 1860', Nature Geoscience, 2, 659–62.
22. J. Galloway et al., 2008: 'Transformation of the Nitrogen Cycle: Recent Trends, Questions, and Potential Solutions', Science, 320, 5878, 889–92.
23. Economic Commission for Europe: Strategies and Policies for Air Pollution Abatement, http://www.unece.org/env/lrtap/ExecutiveBody/2006.Strat.PoliciesReview.E.pdf

24. Galloway et al., cited above.
25. L. Craig et al., 2008: 'Stream restoration strategies for reducing river nitrogen loads', Frontiers in Ecology and the Environment, 6, 10, 529–38.
26. Galloway et al., cited above.
27. G. Hutton et al., 2007: 'Global cost-benefit analysis of water supply and sanitation interventions', Journal of Water and Health, 05.4, 481–502.
28. P. Vitousek et al., 2009: 'Nutrient Imbalances in Agricultural Development', Science, 324, 5934, 1519–20.
29. P. Sanchez and M. Swaminathan, 2005: 'Hunger in Africa: the link between unhealthy people and unhealthy soils', Lancet, 365, 442–4.
30. Smil, cited above.
31. J. Hodgson et al., 2010: 'Comparing organic farming and land sparing: optimizing yield and butterfly populations at a landscape scale', Ecology Letters, 13, 11, 1358–67.
32. http://www.york.ac.uk/news-and-events/news/2010/research/organic-farming/
33. D. Gabriel et al., 2010: 'Scale matters: the impact of organic farming on biodiversity at different spatial scales', Ecology Letters, 13, 7, 858–69.
34. http://www.leeds.ac.uk/news/article/802/organic_farming_shows_limited_benefit_to_wildlife
35. J. Burney et al., 2010: 'Greenhouse gas mitigation by agricultural intensification', PNAS, vol. 107, no. 26, 12052–7.
36. T. Garnett et al., 2009: 'Root based approaches to improving nitrogen use efficiency in plants', Plant, Cell & Environment, 32, 9, 1272–83.
37. A. Good et al., 2007: 'Engineering nitrogen use efficiency with alanine aminotransferase', Canadian Journal of Botany, 85, 3, 252–62.
38. E. Brauer and B. Shelp, 2010: 'Nitrogen use efficiency: re-consideration of the bioengineering approach', Botany, 88, 2, 103–9.
39. A. Shrawat et al., 2008: 'Genetic engineering of improved nitrogen use efficiency in rice by the tissue-specific expression of alanine aminotransferase', Plant Biotechnology Journal, 6, 7, 722–32.
40. Y-M. Bi et al., 2009: 'Increased nitrogen-use efficiency in transgenic rice plants over-expressing a nitrogen-responsive early nodulin gene identified from rice expression profiling', Plant, Cell & Environment, 32, 12, 1749–60.
41. http://www.greenpeace.org/international/en/campaigns/agriculture/problem/genetic-engineering/ge-agriculture-and-genetic-pol/
42. M. Lynas, 2008: 'GM won't yield a harvest for the world', The Guardian, http://www.guardian.co.uk/commentisfree/2008/jun/19/gmcrops.food
43. S. Brand, 2010: Whole Earth Discipline: An Ecopragmatist Manifesto, Atlantic Books, p. 125.

44. S. Duke and S. Powles, 2008: 'Glyphosate: a once-in-a-century herbicide', Pest Management Science, 64, 4, 319–25.
45. Y. Devos et al., 2008: 'Environmental impact of herbicide regimes used with genetically modified herbicide-resistant maize', Transgenic Research, 17, 6, 1059–77.
46. R. Bennett et al., 2004: 'Environmental and human health impacts of growing genetically modified herbicide-tolerant sugar beet: a life-cycle assessment', Plant Biotechnology Journal, 2, 4, 273–8.
47. G. Brookes and P. Barfoot, 2006: 'Global impact of biotech crops: Socio-economic and environmental effects in the first ten years of commercial use', AgBioForum, 9, 3, 139–51.
48. Brookes and Barfoot, cited above.
49. Y. Lu et al., 2010: 'Mirid Bug Outbreaks in Multiple Crops Correlated with Wide-Scale Adoption of Bt Cotton in China', Science, 328, 5982, 1151–4.
50. N. Kingsbury, 2009: Hybrid: the history and science of plant breeding, University of Chicago Press, p.417.
51. Brookes and Barfoot, cited above.
52. D. Plett et al., 2010: 'Improved Salinity Tolerance of Rice Through Cell Type-Specific Expression of AtHKT1;1', PLoS ONE, 5, 9, e12571.
53. J. Chow et al., 2010: 'Cost-Effectiveness of "Golden Mustard" for Treating Vitamin A Deficiency in India', PLoS ONE, 5, 8, e12046.
54. 'Green pepper to the rescue of African bananas', IITA press release, 5 August 2010.
55. M. Moloney and J. Peacock, 2005: 'Plant biotechnology', Current Opinion in Plant Biology 8, 2, 163–4.
56. Committee on the Impact of Biotechnology on Farm-Level Economics and Sustainability, 2010: Impact of Genetically Engineered Crops on Farm Sustainability in the United States, National Research Council.
57. M. Enserink, 2008: 'Tough Lessons From Golden Rice', Science, 320, 5875, 468–71.
58. V. Gewin, 2010: 'Food: An underground revolution', Nature, 466, 552–3.

5: The Land Use Boundary

1. E. Sanderson, et al., 2002: 'The Human Footprint and the Last of the Wild', BioScience, 52, 10, 891–904.
2. J. Lalo, 1987: 'The problem of road kill', American Forests, 93, 50–2.
3. R. Forman and R. Deblinger, 2000: 'The ecological road-effect zone of a Massachusetts (USA) suburban highway', Conservation Biology, 14, 1, 36–46.

4. See J. Diamond, 2006: Collapse: How societies choose to fail or survive, Penguin, 616 pp., for a seminal treatment of the issue.
5. M. Williams, 2006: Deforesting the Earth: From Prehistory to Global Crisis, University of Chicago Press, p. 117.
6. M. Heckenberger et al., 2007: 'The legacy of cultural landscapes in the Brazilian Amazon: implications for biodiversity', Phil. Trans. R. Soc. B, 362, 197–208.
7. P. Kareiva et al., 2007: 'Domesticated Nature: Shaping Landscapes and Ecosystems for Human Welfare', Science, 316, 5833, 1866–9.
8. J. Berger, 2007: 'Fear, human shields and the redistribution of prey and predators in protected areas', Biology Letters, 3, 6, 620–3.
9. M. Ridley, 2009: The Rational Optimist: How prosperity evolves, Fourth Estate, 448 pp.
10. P. Snyder et al., 2004: 'Evaluating the influence of different vegetation biomes on the global climate', Climate Dynamics, 23, 279–302.
11. E. Ellis and N. Ramankutty, 2008: 'Putting people in the map: anthropogenic biomes of the world', Front Ecol Environ, 6, 8, 439–47.
12. H. Haberl et al., 2007: 'Quantifying and mapping the human appropriation of net primary production in earth's terrestrial ecosystems', PNAS, 104, 31, 12942–7.
13. M. Imhoff et al., 2004: 'Global patterns in human consumption of net primary production', Nature, 429, 870–3.
14. S. Kéfi, 2007: 'Spatial vegetation patterns and imminent desertification in Mediterranean arid ecosystems', Nature, 449, 213–17.
15. J. Rockström et al., 2009: 'Planetary Boundaries: Exploring the Safe Operating Space for Humanity', Ecology and Society, 14, 2, 32.
16. J. Ervin, 2003: 'Protected Area Assessments in Perspective', BioScience, 53, 9, 819–22.
17. C. van Schaik et al., 1997: 'The silent crisis: The state of rain forest nature preserves', pp. 64–89 in R. Kramer, C.P. van Schaik and J. Johnson, eds, The Last Stand: Protected Areas and the Defense of Tropical Biodiversity, New York: Oxford University Press.
18. R. Butler, 2010: 'Brazil's Amazon deforestation rate falls to lowest on record', mongabay.com
19. D. Nepstad et al., 2006: 'Inhibition of Amazon deforestation and fire by parks and indigenous lands', Conservation Biology, 20, 1, 65–73.
20. R. Walker et al., 2008: 'Protecting the Amazon with protected areas', PNAS, 106, 26, 10582–6.
21. 'Mato Grosso moves to strip protection of the Amazon rainforest', mongabay.com, 7 November 2010.

22. N. Myers et al., 2000: 'Biodiversity hotspots for conservation priorities', Nature, 403, 853–8.
23. A. Rodrigues et al., 2004: 'Effectiveness of the global protected area network in representing species diversity', Nature, 428, 640–3.
24. R. Mittermeier et al., 2003: 'Wilderness and biodiversity conservation', PNAS, 100, 18, 10309–13.
25. Mittermeier et al.
26. C. Abbott and T. Gardner, 2010: 'U.S. Ethanol Industry Faces Subsidy Battle Next Year', Reuters, 20 December 2010.
27. For the latest see http://www.greenpeace.org.uk/tags/indonesia
28. L. Koh and D. Wilcove, 2008: 'Is oil palm agriculture really destroying tropical biodiversity?', Conservation Letters, 1, 2, 60–4.
29. F. Danielsen et al., 2009: 'Biofuel plantations on forested lands: double jeopardy for biodiversity and climate', Conservation Biology, 23, 2, 348–58.
30. M. Bustamante et al., 2009: 'Chapter 16: What are the Final Land Limits?', in Biofuels: Environmental Consequences and Interactions with Changing Land Use, Proceedings of the Scientific Committee on Problems of the Environment (SCOPE) International, http://cip.cornell.edu/scope/1245782016
31. Myself included, once again. Though I have not to date used air transport exclusively for holidays for over a decade, I hardly need to, given that I work part-time for the President of the Maldives! I have also flown to climate-change negotiations and other climate-related conferences and meetings in various parts of the world.
32. For a primer on aviation and biofuels from the perspective of the industry see http://www.enviro.aero/Content/Upload/File/BeginnersGuide_Biofuels_WebRes.pdf
33. R. Scott, 2010: 'BA agrees deal for UK jet biofuel plant', BBC News, 15 February 2010, http://news.bbc.co.uk/1/hi/8515620.stm
34. V. Fthenakis and H. Kim, 2009: 'Land use and electricity generation: A life-cycle analysis', Renewable and Sustainable Energy Reviews, 13, 6–7, 1465–74.
35. See http://www.rmi.org/rmi/Library/2009-09_FourNuclearMyths for Lovins's arguments against Stewart Brand in particular.
36. K. Smallwood and C. Thelander, 2008: 'Bird Mortality in the Altamont Pass Wind Resource Area, California', Journal of Wildlife Management, 72, 1, 215–23.
37. Mike Norris, 2010: 'Death of large birds concerns naturalist', The Whig-Standard, Kingston, Ontario.
38. R. Barclay et al., 2007: 'Variation in bat and bird fatalities at wind energy facilities: assessing the effects of rotor size and tower height', Canadian Journal of Zoology, 85, 3, 381–7.

39. J. Boyles et al., 2011: 'Economic importance of bats in agriculture', Science, 332, 41–2.
40. T. Kunz et al., 2007: 'Ecological impacts of wind energy development on bats: questions, research needs, and hypotheses', Frontiers in Ecology and the Environment, 5, 315–24.
41. For more see http://www.penycymoeddwindfarm.info/
42. C. Hambler et al., 2010: 'Extinction rates, extinction-prone habitats, and indicator groups in Britain and at larger scales', Biological Conservation, in press. See press release on http://www.ox.ac.uk/media/news_stories/2010/101005.html
43. J. Pethick et al., 2009: 'Nature conservation implications of a Severn tidal barrage – A preliminary assessment of geomorphological change', Journal for Nature Conservation, 17, 4, 183–98.
44. C. Hambler, 'Severn plan leads to mass extinction', The Independent, 30 January 2009.
45. C. Hambler, 2004: Conservation, Cambridge University Press.
46. 'Severn barrage ditched as new nuclear plants get green light', The Guardian, 18 October 2010.
47. P. Denholm and R. Margolis, 2008: 'Land-use requirements and the per-capita solar footprint for photovoltaic generation in the United States', Energy Policy, 36, 9, 3531–43.
48. S. McBride, 2011: 'Special Report: With Solar Power, It's Green vs. Green', Reuters, 6 January 2011.
49. R. McDonald et al., 2009: 'Energy Sprawl or Energy Efficiency: Climate Policy Impacts on Natural Habitat for the United States of America', PLoS ONE, 4, 8.
50. J. Larsen and M. Guillemette, 2007: 'Effects of wind turbines on flight behaviour of wintering common eiders: implications for habitat use and collision risk', Journal of Applied Ecology.
51. http://www.scotland.gov.uk/News/Releases/2010/09/23134359
52. See the MOU at https://www.entsoe.eu/fileadmin/user_upload/_library/news/MoU_North_Seas_Grid/101203_MoU_of_the_North_Seas_Countries_Offshore_Grid_Initiative.pdf
53. H. Bronstein, 2010: 'Ecuador Passes The Hat For Amazon Protection Plan', Reuters, 16 September 2010.
54. S. Creagh, 2010: 'Indonesia Puts Moratorium On New Forest Clearing', Reuters, 28 May 2010.
55. 'Norway and UK commit GBP 100 million to Congo rainforest fund', http://www.norway.org/ARCHIVE/policy/environment/rainforest/
56. G. van der Werf et al., 2009: 'CO_2 emissions from forest loss', Nature Geoscience, 2, 737–8.

57. See http://www.foei.org/en/media/archive/2008/forest-carbon-trading-exposed
58. Friends of the Earth, 2008: 'REDD myths: a critical review of proposed mechanisms to reduce emissions from deforestation and degradation in developing countries', http://www.foe.co.uk/resource/briefing_notes/redd_myths.pdf
59. Friends of the Earth International, 2010: 'Redd: the realities in black and white', November 2010.
60. UNFPA, 2007: State of the World Population 2007 – Unleashing the potential of urban growth, http://www.unfpa.org/swp/2007/presskit/pdf/sowp2007_eng.pdf
61. National Geographic magazine, which is featuring a year-long special on population in 2011, is to thank for this stat.
62. R. Kates and T. Parris, 2003: 'Long-term trends and a sustainability transition', PNAS, 100, 14, 8062–7.
63. Kates and Parris, cited above.
64. Population Division of the Department of Economic and Social Affairs of the United Nations Secretariat, World Population Prospects: The 2008 Revision, http://esa.un.org/unpp
65. UNFPA, 2007: State of the World Population 2007 – Unleashing the potential of urban growth, http://www.unfpa.org/swp/2007/presskit/pdf/sowp2007_eng.pdf
66. A. Lugo and E. Helmer, 2004: 'Emerging forests on abandoned land: Puerto Rico's new forests', Forest Ecology and Management, 190, 2–3, 145–61.
67. T. Aide and H. Grau, 2004: 'Globalization, Migration, and Latin American Ecosystems', Science, 305, 5692, 1915–16.
68. Aide and Grau, cited above.
69. P. Meyfroidt and E. Lambin, 2008: 'The causes of the reforestation in Vietnam', Land Use Policy, 25, 2, 182–97.
70. S. Wright and H. Muller-Landau, 2006: 'The Future of Tropical Forest Species', Biotropica, 38, 3, 287–301.
71. See Figure 2 in the above paper.
72. S. Sloan, 2007: 'Fewer People May Not Mean More Forest for Latin American Forest Frontiers', Biotropica, 39, 4, 443–6.
73. P. Fearnside, 2008: 'Will urbanization cause deforested areas to be abandoned in Brazilian Amazonia?', Environmental Conservation, 35, 197–9.
74. R. Chazdon, 2004: 'Tropical forest recovery: legacies of human impact and natural disturbances', Perspectives in Plant Ecology, Evolution and Systematics, 6, 1–2, 51–71.

6: The Freshwater Boundary

1. S. Postel and B. Richter, 2003: Rivers for Life: Managing Water For People And Nature, Island Press.
2. B. Chao et al., 2008: 'Impact of Artificial Reservoir Water Impoundment on Global Sea Level', Science, 320, 5873, 212–14.
3. B. Chao, 1995: 'Anthropogenic impact on global geodynamics due to reservoir water impoundment', Geophysical Research Letters, 22, 24, 3529–32.
4. Postel and Richter, cited above.
5. L. Gordon et al., 2005: 'Human modification of global water vapor flows from the land surface', PNAS, 102, 21, 7612–17.
6. D. Gerten et al., 2008: 'Causes of change in 20th century global river discharge', Geophysical Research Letters, 35, L20405.
7. F. Saeed et al., 2009: 'Impact of irrigation on the South Asian summer monsoon', Geophysical Research Letters, 36, L20711.
8. G. Hutton and L. Haller, 2004: Evaluation of the Costs and Benefits of Water and Sanitation Improvements at the Global Level, World Health Organisation, http://www.who.int/water_sanitation_health/wsh0404.pdf
9. Chapter 7 of Third UN World Water Development Report, 2009, http://www.unesco.org/water/wwap/wwdr/wwdr3/tableofcontents.shtml
10. Postel and Richter, cited above.
11. C. Taylor, 2010: 'Feedbacks on convection from an African wetland', Geophysical Research Letters, 37, L05406.
12. IUCN, 2008: Freshwater Biodiversity – A hidden resource under threat, http://intranet.iucn.org/webfiles/doc/SpeciesProg/FBU/IUCN_WCC_Freshwater_Factsheet.pdf
13. IUCN, 2010: 'New book celebrates aquatic life in Africa's Okavango Delta', News Story, 4 February 2010.
14. IUCN, 2009: IUCN List of Threatened Species, 2009 Update, Freshwater Fish Facts, http://cmsdata.iucn.org/downloads/more_facts_on_freshwater_fish.pdf
15. Postel and Richter, cited above.
16. G. Miller, 2010: 'In Central California, Coho Salmon Are on the Brink', Science, 327, 5965, 512–13.
17. T. Sun et al., 2008: 'Critical Environmental Flows to Support Integrated Ecological Objectives for the Yellow River Estuary, China', Water Resources Management, 22, 973–89.
18. S. Ghazleh et al., 2009: 'Water input requirements of the rapidly shrinking Dead Sea', Naturwissenschaften, 96, 637–43.
19. Postel and Richter, cited above, p. 125.

20. J. Olden and R. Naiman, 2010: 'Incorporating thermal regimes into environmental flows assessments: modifying dam operations to restore freshwater ecosystem integrity', Freshwater Biology, 55, 86–107.
21. M. Meybeck, 2003: 'Global analysis of river systems: from Earth system controls to Anthropocene syndromes', Phil. Trans. R. Soc. Lond. B, 358, 1935–55.
22. J. Rockström et al., 2009: 'Planetary Boundaries: Exploring the Safe Operating Space for Humanity', Ecology and Society, 14, 2, 32.
23. See http://www.hydroreform.org/sites/www.hydroreform.org/files/RESTORE-for-web.pdf
24. For progress on these and other dam removal projects see www.americanrivers.org
25. See http://www.nature.org/initiatives/freshwater/partnership/
26. Environmental Justice Foundation: 'The Aral Sea Crisis', http://www.ejfoundation.org/page146.html
27. http://www.cottoncampaign.org/frequently-asked-questions/
28. P. Micklin, 2009: 'The Future of the Aral Sea: is the glass half full or half empty', Presentation to the October 2009 St Petersburg Conference 'Aral: Past, Present and Future – Two Centuries of the Aral Sea Investigations', http://www.zin.ru/conferences/Aral2009/ppt/Micklin.pdf
29. International Hydropower Association: 'Hydro's Contribution', http://www.hydropower.org/downloads/F1_The_Contribution_of_Hydropower.pdf
30. International Rivers Network factsheet, 'China's Three Gorges Dam: A model of the past', http://www.internationalrivers.org/files/3Gorges_FINAL.pdf
31. See http://www.fhc.co.uk/dinorwig.htm
32. 'Shoaiba Desalination Plant, Saudi Arabia', on water-technology.net, http://www.water-technology.net/projects/shuaiba/
33. E. Sanchez-Gomez et al., 2009: 'Future changes in the Mediterranean water budget projected by an ensemble of regional climate models', Geophysical Research Letters, 36, L21401.
34. J. Sheffield and E. Wood, 2007: 'Projected changes in drought occurrence under future global warming from multi-model, multi-scenario, IPCC AR4 simulations', Climate Dynamics, 31, 1, 79–105.
35. S. Rauscher et al., 2008: 'Future changes in snowmelt-driven runoff timing over the western US', Geophysical Research Letters, 35, L16703.
36. D. Renault (2003) 'Value of virtual water in food: Principles and virtues', in http://www.waterfootprint.org/Reports/Report12.pdf
37. Reuters, 2010: 'Cut Back On Farming To Save Water, UAE Told', 9 February 2011.
38. T. Oki et al., in http://www.waterfootprint.org/Reports/Report12.pdf

39. Virtual water trade: A quantification of virtual water flows between nations in relation to international crop trade, A.Y. Hoekstra and P.Q. Hung, in http://www.waterfootprint.org/Reports/Report12.pdf
40. K. Gassner et al., 2009: Does private sector participation improve performance in electricity and water distribution?, World Bank Publications, 98 pp.
41. S. Galiani et al., 2005: 'Water for Life: The Impact of the Privatization of Water Services on Child Mortality', Journal of Political Economy, 113, 1, 83.
42. A. Rabinovitch, 2010: 'Arid Israel Recycles Waste Water On Grand Scale', Reuters, 15 November 2010.
43. D. Renault, (2003) 'Value of virtual water in food: Principles and virtues', in http://www.waterfootprint.org/Reports/Report12.pdf

7: The Toxics Boundary

1. D. Barnes et al., 2009: 'Accumulation and fragmentation of plastic debris in global environments', Philos. Trans. R. Soc. Lond. B. Biol. Sci., 364, 1526, 1985–98.
2. K. Weiss, 2006: 'Plague of Plastic Chokes the Seas', Los Angeles Times, http://www.latimes.com/news/printedition/la-me-ocean2aug02,0,5274274,full.story
3. http://www.zsl.org/science/news/extinct-in-ten-years-vultures-decline-quicker-than-the-dodo,450,NS.html
4. M. Milnes and L. Guillette Jnr, 2008: 'Alligator Tales: New Lessons about Environmental Contaminants from a Sentinel Species', BioScience, 58, 11, 1027–36.
5. C. Tyler and S. Jobling, 2008: 'Roach, Sex, and Gender-Bending Chemicals: The Feminization of Wild Fish in English Rivers', BioScience, 58, 11, 1051–9.
6. G. Bryan et al., 1986: 'The Decline of the Gastropod Nucella Lapillus Around South-West England: Evidence for the Effect of Tributyltin from Antifouling Paints', Journal of the Marine Biological Association of the United Kingdom, 66, 611–40.
7. P. Gibbs and G. Bryan, 1986: 'Reproductive Failure in Populations of the Dog-Whelk, Nucella Lapillus, Caused by Imposex Induced by Tributyltin from Antifouling Paints', Journal of the Marine Biological Association of the United Kingdom, 66, 767–77.
8. E. Thurman and A. Cromwell, 2000: 'Atmospheric Transport, Deposition, and Fate of Triazine Herbicides and Their Metabolites in Pristine Areas at Isle Royale National Park', Environmental Science and Technology, 34, 15, 3079–85.

9. C. Miljeteig et al., 2009: 'High Levels of Contaminants in Ivory Gull Pagophila eburnea Eggs from the Russian and Norwegian Arctic', Environmental Science and Technology, 43, 14, 5521–8.
10. BirdLife International, 2008: Pagophila eburnea. In: IUCN 2010. IUCN Red List of Threatened Species.
11. N. Michelutti et al., 2009: 'Seabird-driven shifts in Arctic pond ecosystems', Proc. R. Soc. B, 276, 1656, 591–6.
12. E. Dewailly et al., 1993: 'Inuit exposure to organochlorines through the aquatic food chain in arctic québec', Environ. Health Perspect., 101, 7, 618–20.
13. J. Van Oostdam et al., 1999: 'Human health implications of environmental contaminants in Arctic Canada: a review', The Science of the Total Environment, 230, 1–3, 1–82.
14. M. Benotti et al., 2009: 'Pharmaceuticals and Endocrine Disrupting Compounds in U.S. Drinking Water', Environmental Science and Technology, 43, 3, 597–603.
15. H. Fisch et al., 1996: 'Semen analyses in 1283 men from the United States over a 25-year period: no decline in quality', Fertility and Sterility, 65, 1009–14.
16. H. Fisch, 2008: 'Declining worldwide sperm counts: Disproving a myth', Urologic Clinics of North America, 35, 137–47.
17. M.-H. Wang et al., 2008: 'Endocrine Disruptors, Genital Development, and Hypospadias', Journal of Andrology, 29, 499–505.
18. H. Fisch et al., 2010: 'Rising hypospadias rates: Disproving a myth', Journal of Pediatric Urology, 6, 1, 37–9.
19. M. López-Cervantes et al., 2004: 'Dichlorodiphenyldichloroethane burden and breast cancer risk: A meta-analysis of the epidemiologic evidence', Environmental Health Perspectives, 112, 2, 207–14.
20. M. Gammon et al., 2002: 'Environmental toxins and breast cancer on Long Island. II. Organochlorine compound levels in blood', Cancer Epidemiology Biomarkers and Prevention, 11, 8, 686–97.
21. S. Safe, 2004: 'Endocrine disruptors and human health: is there a problem', Toxicology, 205, 1–2, 3–10.
22. A. Blaustein and P. Johnson, 2003: 'The complexity of deformed amphibians', Front. Ecol. Environ., 1, 2, 87–94.
23. http://www.epa.gov/safewater/contaminants/basicinformation/atrazine.html
24. T. Hayes et al., 2010: 'Atrazine induces complete feminization and chemical castration in male African clawed frogs (Xenopus laevis)', PNAS, 107, 10, 4612–17.
25. W. Kloas, 2009: 'Does Atrazine Influence Larval Development and Sexual Differentiation in Xenopus laevis?', Toxicological Sciences, 107, 2, 376–84.

26. R. Renner, 2008: 'Atrazine effects in Xenopus aren't reproducible', Environmental Science and Technology, 42, 10, 3491–3.
27. D. Wake and V. Vrendenburg, 2008: 'Are we in the midst of the sixth mass extinction? A view from the world of amphibians', PNAS, 105, 11466–73.
28. V. Vrendenburg et al., 2010: 'Dynamics of an emerging disease drive large-scale amphibian population extinctions', PNAS, 107, 21, 9689–94.
29. http://chm.pops.int/Convention/Media/Pressreleases/COP4Geneva9May2009/tabid/542/language/en-US/Default.aspx
30. http://guidance.echa.europa.eu/about_reach_en.htm
31. N. Gilbert, 2009: 'Chemical-safety costs uncertain', Nature, 460, 1065.
32. Editorial, 2010: 'The weight of evidence', Nature, 464, 1103–4.
33. http://www.greenpeace.org.uk/toxics/chemicalhome
34. http://www.euractiv.com/en/energy/solar-industry-divided-over-toxic-substances-law-news-468176
35. http://www.ntsa.eu/
36. http://www.greenpeace.org/international/en/campaigns/peace/abolish-nuclear-weapons/the-damage/
37. S. Watson et al., 'Ionising radiation exposure of the UK population: 2005 review', HPA, http://www.hpa.org.uk/web/HPAwebFile/HPAweb_C/1194947389360
38. M. Ghiassi-nejad et al., 2002: 'Very high background radiation areas of Ramsar, Iran: preliminary biological studies', Health Physics, 82, 1, 87–93.
39. Z. Tao et al., 2000: 'Cancer Mortality in the High Background Radiation Areas of Yangjiang, China during the Period between 1979 and 1995', Journal of Radiation Research, 41, S1–S7.
40. World Health Organisation, 2009: Radon and Cancer, Fact Sheet no. 291.
41. D. Laurier et al., 2002: 'Risk of Childhood Leukaemia in the Vicinity of Nuclear Installations: Findings and Recent Controversies', Acta Oncologica, 41, 1 14–24.
42. P. Kaatsch et al., 2008: 'Leukaemia in young children living in the vicinity of German nuclear power plants', International Journal of Cancer, 122, 4, 721–6.
43. P. Cook-Mozaffari et al., 1989: 'Cancer near potential sites of nuclear installations', The Lancet, 334, 8672, 1145–7.
44. F. Zölzer, 2010: 'Childhood leukaemia in the vicinity of German nuclear power plants – some missing links', Journal of Applied Biomedicine, 8, 1–6.
45. 'Chernobyl deaths "up to 66,000"', Scotland on Sunday, 23 April 2006, http://scotlandonsunday.scotsman.com/nuclearincidents/Chernobyl-deaths-up-to-66000.2769657.jp
46. http://www.iaea.org/Publications/Booklets/Chernobyl/chernobyl.pdf
47. http://www.unscear.org/docs/reports/2000/Volume%20II_Effects/AnnexJ_pages%20451-566.pdf

48. UNSCEAR, as cited above.
49. http://www.greenpeace.org/international/en/news/features/chernobyl-deaths-180406/
50. http://www.greenpeace.org/international/Global/international/planet-2/report/2006/4/chernobylhealthreport.pdf
51. Personal communication from the Ukrainian government. For comparison of radiation doses received for medical purposes, see for example http://www.hps.org/hpspublications/articles/dosesfrommedicalradiation.html
52. http://www-pub.iaea.org/mtcd/publications/pdf/pub1239_web.pdf
53. R. Chesser and R. Baker, 2006: 'Growing up with Chernobyl', American Scientist, 94, 542–9.
54. http://www-pub.iaea.org/mtcd/publications/pdf/pub1239_web.pdf
55. See A. Møller and T. Mousseau, 2009: 'Reduced abundance of insects and spiders linked to radiation at Chernobyl 20 years after the accident', Biol. Lett., 5, 3, 356–9, and 2010: 'Efficiency of bio-indicators for low-level radiation under field conditions', Ecological Indicators, in press.
56. See for example, J. Smith, 2008: 'Is Chernobyl radiation really causing negative individual and population-level effects on barn swallows?', Biol. Lett., 23 February 2008 vol. 4, no. 1, 63–4.
57. B. Borrell, 2007: 'A Fluctuating Reality: Accused of fraud, Anders Pape Möller has traveled from superstar evolutionary biologist to pariah', The Scientist, 21, 1, p. 26.
58. A. Møller et al., 2011: 'Chernobyl Birds Have Smaller Brains', PLoS One, 6, 2, e16862.
59. Next Big Future blog, 2011: 'Deaths per TWH by energy source', http://nextbigfuture.com/2011/03/deaths-per-twh-by-energy-source.html
60. N. Starfelt and C. Wikdahl, undated: 'Economic analysis of various options of electricity generation – taking into account the health and environmental effects', http://manhaz.cyf.gov.pl/manhaz/strona_konferencja_EAE-2001/15%20-%20Polenp~1.pdf
61. A. Madrigal, 2011: '25 Other Energy Disasters From the Last Year', The Atlantic, http://www.theatlantic.com/technology/archive/2011/03/25-other-energy-disasters-from-the-last-year/72814/
62. http://www.foe.co.uk/resource/briefing_notes/nuclear_not_a_solution.pdf
63. http://www.foe.co.uk/resource/press_releases/nuclear_power_a_dangerous_26032008.html
64. See Figure 5 and discussion, N. Chapman and C. Curtis, 2006: 'Confidence in the Safe Geological Disposal of Radioactive Waste', post-conference report, http://www.geolsoc.org.uk/webdav/site/GSL/shared/pdfs/our%20views/radwaste/Chapman%20and%20Curtis%20report.pdf
65. http://www.world-nucleur.org/info/inf04.html

66. Assuming that each station avoids the emission from a comparable 1-GW coal plant of 6 million tonnes of carbon dioxide per year.
67. http://www.independent.co.uk/environment/green-living/nuclear-power-yes-please-1629327.html
68. G. Monbiot, 2011: 'The double standards of green anti-nuclear opponents', The Guardian, http://www.guardian.co.uk/environment/georgemonbiot/2011/mar/31/double-standards-nuclear
69. Reuters, 2010: 'Plans For 150 Coal Plants Scrapped: Sierra Club', 7 February 2011.
70. Greenpeace International, 2011: 'Battle of the grids', http://www.greenpeace.org/international/en/publications/reports/Battle-of-the-grids/

8: The Aerosols Boundary

1. K. Wang et al., 'Clear Sky Visibility Has Decreased over Land Globally from 1973 to 2007', Science, 323, 5920, 1468–70.
2. D. Hofmann et al., 2009: 'Increase in background stratospheric aerosol observed with lidar at Mauna Loa Observatory and Boulder, Colorado', Geophysical Research Letters, 36, L15808.
3. W. Randel et al., 2010: 'Asian Monsoon Transport of Pollution to the Stratosphere', Science, 328, 5978, 611–13.
4. G. Carmichael et al., 2009: 'Asian Aerosols: Current and Year 2030 Distributions and Implications to Human Health and Regional Climate Change', Environmental Science & Technology, 43, 15, 5811–17.
5. R. Stone, 2008: 'Beijing's Marathon Run to Clean Foul Air Nears Finish Line', Science, 321, 5889, 636–7.
6. http://www.publications.parliament.uk/pa/cm200910/cmselect/cmenvaud/229/22905.htm
7. http://www.stateoftheair.org/2010/key-findings/
8. A. Appatova, et al., 2008: 'Proximal exposure of public schools and students to major roadways: a nationwide US survey', Journal of Environmental Planning and Management, 51, 5, 631–46.
9. H. Kawase et al., 2010: 'Physical mechanism of long-term drying trend over tropical North Africa', Geophysical Research Letters, 37, L09706.
10. I. Held et al., 2005: 'Simulation of Sahel drought in the 20th and 21st centuries', PNAS, 102, 50, 17891–6.
11. V. Ramanathan et al., 2005: 'Atmospheric brown clouds: Impacts on South Asian climate and hydrological cycle', PNAS, 102, 15, 5326–33.
12. M. Wild et al., 2008: 'Combined surface solar brightening and increasing greenhouse effect support recent intensification of the global land-based hydrological cycle', Geophysical Research Letters, 35, L17706.

13. V. Ramanathan and G. Carmichael, 2008: 'Global and regional climate changes due to black carbon', Nature Geoscience, 1, 221–7.
14. G. Meehl et al., 2008: 'Effects of Black Carbon Aerosols on the Indian Monsoon', Journal of Climate, 21, 2869–82.
15. Q. Xu, 2001: 'Abrupt Change of the Mid-Summer Climate in Central East China by the Influence of Atmospheric Pollution', Atmospheric Environment, 35, 30, 5029–40.
16. A. Stohl, 2006: 'Characteristics of atmospheric transport into the Arctic troposphere', J. Geophys. Res., 111, D11306.
17. C. Warneke et al., 2010: 'An important contribution to springtime Arctic aerosol from biomass burning in Russia', Geophysical Research Letters, 37, L01801.
18. J. Hansen and L. Nazarenko, 2004: 'Soot climate forcing via snow and ice albedos', PNAS, 101, 2, 423–8.
19. M. Fiebig et al., 2009: 'Tracing biomass burning aerosol from South America to Troll Research Station, Antarctica', Geophysical Research Letters, 36, L14815.
20. H. Venzac et al., 2008: 'High frequency new particle formation in the Himalayas', PNAS, 105, 41, 15666–71.
21. S. Marcq et al., 2010: 'Aerosol optical properties and radiative forcing in the high Himalaya based on measurements at the Nepal Climate Observatory – pyramid site (5100 m a.s.l)', Atmos. Chem. Phys. Discuss., 10, 5627–63.
22. B. Xu et al., 2009: 'Black soot and the survival of Tibetan glaciers', PNAS, 106, 52, 22114–18.
23. M. Jacobson, 2010: 'Short-term effects of controlling fossil-fuel soot, biofuel soot and gases, and methane on climate, Arctic ice, and air pollution health', J. Geophys. Res., 115, D14209.
24. Hansen and Nazarenko, cited above.
25. J. McNally, 2010: 'Best hope for saving Arctic sea ice is cutting soot emissions, says Stanford researcher', Stanford Report, 28 July 2010, http://news.stanford.edu/news/2010/july/soot-emissions-ice-072810.html
26. See Figure 2 in T. Bond and H. Sun, 2005: 'Can Reducing Black Carbon Emissions Counteract Global Warming?', Environmental Science and Technology, 39, 16, 5921–6.
27. W. Chameides and M. Bergin, 2002: 'Soot Takes Center Stage', Science, 297, 5590, 2214–15.
28. J. Corbett et al., 2007: 'Mortality from Ship Emissions: A Global Assessment', Environmental Science and Technology, 41, 24, 8512–18.
29. Here I am comparing with future predicted deaths from Chernobyl (rather than with the actual current death toll of 50 or so), making a reasonable guess of 4,000 additional fatal cancer cases overall in the affected

populations over the next few decades. Of course, we will never know for sure.

30. F. Kan and R. Fabi, 2011: 'Analysis: Refiners Threaten Anti-Pollution Efforts In Shipping', Reuters, 18 January 2011.
31. Q. Zhang et al., 2009: 'Asian emissions in 2006 for the NASA INTEX-B mission', Atmospheric Chemistry and Physics, 9, 5131–53.
32. Ramanathan and Carmichael, cited above.
33. Bond and Sun, cited above, table 1.
34. M. Frondel et al., 2009: 'Economic Impacts from the Promotion of Renewable Energy Technologies – The German Experience', Ruhr Economic Papers #156, http://repec.rwi-essen.de/files/REP_09_156.pdf
35. Carbon Trade Watch, 2007: The Carbon Neutral Myth – Offset Indulgences for your Climate Sins, http://www.carbontradewatch.org/pubs/carbon_neutral_myth.pdf
36. A. Ma'anit, 2007: 'If you go down to the woods today …', New Internationalist, 391, http://www.newint.org/features/2006/07/01/keynote/
37. World Bank, 2010: State and Trends of the Carbon Market 2010, Carbon Finance at the World Bank.
38. S. Rasool and S. Schneider, 1971: 'Atmospheric Carbon Dioxide and Aerosols: Effects of Large Increases on Global Climate', Science, 173, 3992, 138–41.
39. V. Vestreng et al., 2007: 'Twenty-five years of continuous sulphur dioxide emission reduction in Europe', Atmospheric Chemistry and Physics, 7, 3663–81.
40. M. Wild, 2009: 'Global dimming and brightening: A review', J. Geophys. Res., 114, D00D16.
41. M. Wild et al., 2007: 'Impact of global dimming and brightening on global warming', Geophysical Research Letters, 34, L04702.
42. P. Crutzen, 2006: 'Albedo Enhancement by Stratospheric Sulfur Injections: A Contribution to Resolve a Policy Dilemma?', Climatic Change, 77, 3–4, 211–20.
43. A. Robock et al., 2010: 'A Test for Geoengineering?', Science, 327, 530–1.
44. K. Trenberth and A. Dai, 2007: 'Effects of Mount Pinatubo volcanic eruption on the hydrological cycle as an analog of geoengineering', Geophysical Research Letters, 34, L15702.
45. O. Morton, 2007: 'Climate change: Is this what it takes to save the world?', Nature, 447, 132–6.
46. But see, for example, A. Robock, 2008: '20 reasons why geoengineering may be a bad idea', Bulletin of Atomic Scientists, 64, 2, 14–18.
47. Crutzen, cited above.

9: The Ocean Acidification Boundary

1. S.C. Doney et al., 2009: 'Ocean acidification: a critical emerging problem for the ocean sciences', Oceanography, 22,4, 16–25.
2. U. Riebesell et al., 2009: 'Sensitivities of marine carbon fluxes to ocean change', PNAS, 106, 49 20602–9.
3. Here things get less simple. Carbonic acid dissociates into bicarbonate and hydrogen ions (protons). Most marine organisms use carbonate for their shells, and amounts of carbonate (CO_3^{2-}) tend to be depleted as a result of this process. For a good discussion of the chemistry involved, see R. Feely et al., 2009: 'Ocean acidification: Present conditions and future changes in a high-CO_2 world', Oceanography, 22, 4, 36–47.
4. C. Pelejero et al., 2010: 'Paleo-perspectives on ocean acidification', Trends in Ecology & Evolution, 25, 6, 332–4.
5. Feely et al., cited above.
6. J. Dore et al., 2009: 'Physical and biogeochemical modulation of ocean acidification in the central North Pacific', PNAS, 106, 30, 12235–40.
7. Pelejero et al., cited above.
8. R. Feely et al., 2008: 'Evidence for Upwelling of Corrosive "Acidified" Water onto the Continental Shelf', Science, 320, 1490–2.
9. M. Yamamoto-Kawai et al., 2009: 'Aragonite Undersaturation in the Arctic Ocean: Effects of Ocean Acidification and Sea Ice Melt', Science, 326, 5956, 1098–1100.
10. G. De'ath et al., 2009: 'Declining Coral Calcification on the Great Barrier Reef', Science, 323, 5910, 116–19.
11. R. Bak et al., 2009: 'Coral growth rates revisited after 31 years: What is causing lower extension rates in Acropora palmata?', Bulletin of Marine Science, 84, 3, 287–94.
12. J. Tanzil et al., 2009: 'Decline in skeletal growth of the coral Porites lutea from the Andaman Sea, South Thailand between 1984 and 2005', Coral Reefs, 28, 2, 519–28.
13. K. Anthony et al., 2008: 'Ocean acidification causes bleaching and productivity loss in coral reef builders', PNAS, 105, 45, 17442–6.
14. J. Hall-Spencer et al., 2008: 'Volcanic carbon dioxide vents show ecosystem effects of ocean acidification', Nature, 454, 96–9.
15. D. Manzello et al., 2008: 'Poorly cemented coral reefs of the eastern tropical Pacific: Possible insights into reef development in a high-CO_2 world', PNAS, 105, 30, 10450–5.
16. R. Rodolfo-Metalpa et al., 2010: 'Response of the temperate coral Cladocora caespitosa to mid- and long-term exposure to pCO_2 and temperature levels projected for the year 2100 AD', Biogeosciences, 7, 289–300.

17. J. Silverman et al., 2009: 'Coral reefs may start dissolving when atmospheric CO_2 doubles', Geophysical Research Letters, 36, L05606.
18. Yamamoto-Kawai et al., cited above.
19. C. Hauri et al., 2009: 'Ocean acidification in the California Current System' Oceanography, 22, 4, 60–71.
20. P. Brewer and E. Peltzer, 2009: 'Oceans: Limits to Marine Life', Science, 324, 5925, 347–8.
21. A. Moy, et a.l, 2009: 'Reduced calcification in modern Southern Ocean planktonic foraminifera', Nature Geoscience, 2, 276–80.
22. W. Balch, and P. Utgoff, 2009: 'Potential interactions among ocean acidification, coccolithophores, and the optical properties of seawater', Oceanography, 22, 4, 146–59.
23. See, for example, two reports spanning a decade – U. Riebesell et al., 2000: 'Reduced calcification of marine plankton in response to increased atmospheric CO_2', Nature, 407, 364–7; and M. Muller et al., 2010: 'Effects of long-term high CO_2 exposure on two species of coccolithophores', Biogeosciences, 7, 1109–16.
24. J. Orr,et al., 2005: 'Anthropogenic ocean acidification over the twenty-first century and its impact on calcifying organisms', Nature, 437, 681–6.
25. S. Kawaguchi et al., 2010: 'Will krill fare well under Southern Ocean acidification?', Biology Letters, in press.
26. P. Brewer and K. Hester, 2009: 'Ocean acidification and the increasing transparency of the ocean to low-frequency sound', Oceanography, 22, 4, 86–93.
27. P. Munday et al., 2010: 'Replenishment of fish populations is threatened by ocean acidification', PNAS, 107, 29, 12930–4.
28. F. Gazeau et al., 2010: 'Effect of ocean acidification on the early life stages of the blue mussel (Mytilus edulis)', Biogeosciences Discussions, 7, 2927–47.
29. J. Kleypas and K. Yates, 2009: 'Coral reefs and ocean acidification', Oceanography, 22, 4, 108–17.
30. D. Hutchins et al., 2009: 'Nutrient cycles and marine microbes in a CO2-enriched ocean', Oceanography, 22, 4, 128–45.
31. J. Jackson, 2008: 'Ecological extinction and evolution in the brave new ocean', PNAS, 105, Supplement 1, 11458–65.
32. J. Veron, 2008: 'Mass extinctions and ocean acidification: biological constraints on geological dilemmas', Coral Reefs, 27, 3, 459–72.
33. A. Marzoli et al., 2004: 'Synchrony of the Central Atlantic magmatic province and the Triassic–Jurassic boundary climatic and biotic crisis', Geology, 32, 11, 973–6.

34. M. Hautmann et al., 2008: 'Catastrophic ocean acidification at the Triassic–Jurassic boundary', Neues Jahrbuch für Geologie und Paläontologie – Abhandlungen, 249, 1, 119–27.
35. J. Whiteside et al., 2010: 'Compound-specific carbon isotopes from Earth's largest flood basalt eruptions directly linked to the end-Triassic mass extinction', PNAS, 107, 15, 6721–5.
36. B. van de Schootbrugge et al., 2007: 'End-Triassic calcification crisis and blooms of organic-walled "disaster species"', Palaeogeography, Palaeoclimatology, Palaeoecology, 244, 1–4, 126–41.
37. L. Kump et al., 2009: 'Ocean acidification in deep time', Oceanography, 22, 4, 94–107.
38. M. Medina et al., 2006: 'Naked corals: Skeleton loss in Scleractinia', PNAS, 103, 24, 9096–9100.
39. Kump et al., cited above.
40. R. Kerr, 2010: 'Ocean Acidification Unprecedented, Unsettling', Science, 328, 5985, 1500–1.
41. Pelejero et al., cited above.
42. See http://trillionthtonne.org/ for the latest.
43. W. Kiessling, and C. Simpson, 2010: 'On the potential for ocean acidification to be a general cause of ancient reef crises', Global Change Biology, 17, 1, 56–67.
44. Pelejero et al., cited above.
45. A. Ridgwell and D. Schmidt, D., 2010: 'Past constraints on the vulnerability of marine calcifiers to massive carbon dioxide release', Nature Geoscience, 3, 196–200.
46. O. Hoegh-Guldberg, et al., 2007: 'Coral Reefs Under Rapid Climate Change and Ocean Acidification', Science, 318, 5857, 1737–42.
47. Pelejero et al., cited above.
48. M. Steinacher et al., 2009: 'Imminent ocean acidification in the Arctic projected with the NCAR global coupled carbon cycle-climate model', Biogeosciences, 6, 515–33.
49. J. Veron et al., 2009: 'The coral reef crisis: The critical importance of <350 ppm CO_2', Marine Pollution Bulletin, 58, 10, 1428–36.
50. J. Rockström et al., 2009: 'Planetary Boundaries: Exploring the Safe Operating Space for Humanity', Ecology and Society, 14, 2, 32.
51. Figures from Table 1, J. Guinotte and V. Fabry, 2008: 'Ocean Acidification and Its Potential Effects on Marine Ecosystems', Annals of the New York Academy of Sciences, 1134, Issue 'The Year in Ecology and Conservation Biology 2008', 320–42.
52. R. Schuiling and O. Tickell, 2010: 'Olivine against climate change and ocean acidification', unpublished manuscript communicated by the authors.

53. For a detailed description see http://www.cquestrate.com/the-idea/detailed-description-of-the-idea
54. M. Ridley, 2009: The Rational Optimist, Fourth Estate, p. 340.
55. http://www.nationalreview.com/planet-gore/250180/global-warming-s-corrupt-science-patrick-j-michaels
56. M. Ridley, 2010: 'Who's afraid of acid in the ocean? Not me', The Times, 4 November 2010.
57. See 'Response on behalf of UK Ocean Acidification Research Programme' to Matt Ridley article, on http://www.oceanacidification.org.uk/PDF/Briefing%20note%20on%20Ridley%20article%20-%2019%20Nov.pdf
58. http://wonkroom.thinkprogress.org/climate-zombie-caucus/
59. http://dailycaller.firenetworks.com/001646/dailycaller.com/wp-content/blogs.dir/1/files/Tea-Party-Dec-of-Independence-22410.pdf
60. http://www.greenparty.org.uk/assets/files/reports/the_new_home_front_FINAL.pdf
61. See, for example, some good points made in response in Rob Lyons's piece on Spiked: http://www.spiked-online.com/index.php/site/article/10116/

10: The Ozone Layer Boundary

1. J. McNeill, 2000: Something new under the sun – an environmental history of the twentieth century, Penguin Books, p. 211.
2. P. Crutzen, 1995 Nobel Lecture, http://nobelprize.org/nobel_prizes/chemistry/laureates/1995/crutzen-lecture.pdf
3. J. Lovelock et al., 1973: 'Halogenated Hydrocarbons in and over the Atlantic', Nature, 241, 194–6; the mistake is acknowledged in Lovelock's autobiography Homage to Gaia.
4. M. Molina and F. Rowland, 1974: 'Stratospheric sink for chlorofluoromethanes: chlorine atom-catalysed destruction of ozone', Nature, 249, 5460, 810–12.
5. See Figure 1 in M. Chipperfield, 2009: 'Atmospheric science: Nitrous oxide delays ozone recovery', Nature Geoscience 2, 742–3.
6. See Annual Records table in NASA's Ozone Hole Watch website: http://ozonewatch.gsfc.nasa.gov/meteorology/annual_data.html
7. P. Newman et al., 2009: 'What would have happened to the ozone layer if chlorofluorocarbons (CFCs) had not been regulated?', Atmospheric Chemistry and Physics, 9, 2113–28.
8. E. Parson, 2003: Protecting the Ozone Layer, Science and Strategy, Oxford University Press.

11: Managing the Planet

1. There is a picture of Yu Qingtai on http://beta.thehindu.com/news/article62882.ece?homepage=true He is sometimes incorrectly identified as Su Wei, China's lead negotiator at the Copenhagen talks.
2. http://www.america.gov/st/peacesec-english/2009/July/20090709153615esnamfuak0.3577082.html
3. J. Rockström et al., 2009: 'A safe operating space for humanity', Nature, 461, 472–5.
4. Dr Pieter Tans, NOAA/ESRL (www.esrl.noaa.gov/gmd/ccgg/trends/), current at January 2011.
5. B. O'Neill et al., 2010: 'Global demographic trends and future carbon emissions', PNAS, in press.
6. http://www.jonathonporritt.com/pages/population/
7. http://www.jonathonporritt.com/pages/population/
8. D. Helm, 2008: 'Climate-change policy: why has so little been achieved?', Oxford Review of Economic Policy, 24, 2, 211–38.
9. UK emissions in 2009 were 520 million tonnes, whilst those of Spain and Ireland were 330 and 40 respectively. Figures from the US Energy Information Administration, via http://tonto.eia.doe.gov/cfapps/ipdbproject/iedindex3.cfm?tid=90&pid=44&aid=8&cid=regions,&syid=2005&eyid=2009&unit=MMTCD
10. J. Guo et al., 2010: 'Significant Acidification in Major Chinese Croplands', Science, 327, 5968, 1008–10.
11. J. Ausubel and P. Waggoner, 2008: 'Dematerialization: Variety, caution, and persistence', PNAS, 105, 35, 12774–9.
12. Bloomberg, 2010: 'China Beats U.S. on Renewable-Energy Investor Ranking', 8 September 2010.
13. Global Wind Energy Council, 2011: 'Global wind capacity increases by 22% in 2010 – Asia leads growth', 2 February 2011.

INDEX

aerosols boundary 183–97; sky colour and 183–4; Asian Brown Cloud 184–5, 186; human suffering from air pollution 185–6; hydrological cycle and 186–7; black carbon 188–93; sulphur emissions 193–7; solar radiation management 194–7
Africa: hominids in 27, 28, 29, 34; endangered animals in 32, 36; poverty in 68; solar power in 77; shortage of fertiliser 98, 99; genetic engineering in 106, 107; safe drinking water in 141; climate change in 151, 186; monsoon in 195; growth of economy in 239
agriculture: invention of 22; threatens rainforest 63; nitrogen boundary and 85–109; organic 85–6, 87, 98–101, 118, 119; Green Revolution 92, 98, 141–2, 155, 156; genetic engineering 3–4, 11, 12, 101–9, 119, 214; no-till 105, 119; intensification of 112, 118–19, 136–7; land use 115, 118–19, 122, 126; high-yield 92, 98, 119, 136–7, 141–2; irrigation 140, 141–2; water use 140–4, 148, 149, 152, 153, 155–6; pesticides 104–7, 118, 119, 141–2, 158, 163, 165, 218 *see also under individual pesticide name*
agro-forestry 119
air travel/aviation 80, 123–4, 192, 196
Allen, Myles 84
Alliance for Responsible CFC Policy 221–2
Altamont Pass Wind Resource Area 125–6
Amazon rainforest 62–3, 116–17, 131, 137, 187
Amazon River 111
ammonia production 88–92, 95, 96, 109, 217–18
Amu Darya 148–9
An Appeal to Reason: A cool look at global warming (Lawson) 212
Andes 59, 151
Andreae, Meinrat 195
Antarctic 38, 61–2, 65, 118, 187, 202, 218, 219, 220
Anthropocene 5, 10, 12, 21, 30, 66, 84, 183, 199
aquaculture 119
aragonite 207, 208, 236
Aral Sea 148–9, 152
Archer, David 64
Arctic: plastic waste in 5, 151; habitat destruction in 40; thaw of 56–7, 58, 59, 60, 61–2, 63, 64, 65, 188; tundra 114, 117, 118; toxins accumulate in 160, 171, 187; ocean acidification and 200, 201, 202, 207
Argentina 106, 136, 154
Argonne National Laboratory 81
Asia: tsunami, 2004 7; *Homo neanderthalensis* 27; animal extinction and 32, 143; poverty in 68; wind power in 74; nitrate pollution in 97; genetically engineered crops in 107, 108; protected areas in 118; urbanization 137; storm surges in 146; aerosol pollution in 183, 184–7, 190–1, 195
Asian Brown Cloud 184–5, 186
Aswan Dam 73, 94
Atlantic Ocean: global warming destabilises circulation of 62
Atlantic Wind Connection 74
Atomic Energy Agency 176
atrazine 159–60, 161, 163–4
Australia: extinction in 58–9; climate change in 58–9; solar power potential 73–4, 128; virtual water 153; Great Barrier Reef 200–1
Australoptihecus 27
Austria 70, 180

Baker, Robert 175
Baltic Sea 92
'Bank of Natural Capital' 48
BASF 89, 90
Berlins, Marcel 42
Better Place 123
'BioBanking' scheme, Malua 49, 50
'Biodiversity Conservation Certificates' 49–50
biodiversity loss 12, 19–20; boundary 30–51; accounting systems for 30–1; 'biodiversity credits' 49; extinction and 31–3; Pleistocene overkill 33–4; eliminating alien species form islands 41; and the earth system 41–6; keystone predators 43; habitat loss 44–5; 'paper parks' 46; valuing of natural systems 46–9; global 'tipping point' 45; planetary boundary on 45; offsets 50; protection measures 50–1; biodiversity 'hotspots' 117
biofuels 4, 46, 80, 95, 121, 122, 123, 124, 129

biomass 5, 20, 70, 76, 81, 86, 99, 114, 125, 175, 185
BioScience 63
biosphere: monetary value of 47
black carbon 57, 188–93
Borneo 32, 46, 49, 122
Bosch, Carl 88, 89, 90
Boyles, Justin 126
BP 227
brain: evolution of 5, 6, 25, 26, 27–8
Brand, Stewart 22, 181
Brazil 36, 62, 68, 71, 106, 115, 116, 131, 204, 225, 232
British Airways 124
Broecker, Wally 65
Brown, Gordon 232
Bush, George W. 225

cadmium 160, 166, 167
calcium carbonate 18, 199, 200, 202, 204, 205, 206, 207, 208
Calcutta 145–6
Cambrian explosion 23–4
Canada 102, 118, 119, 147, 160, 200
Cancun, UN climate-change meeting, 2010 133, 234, 242
'cap and trade' programmes 79, 84, 155, 193
carbon: cycle 5, 9, 15–16, 17–18, 19, 20, 21, 22, 23, 29, 198–9, 209; offsetting/markets 1, 11, 50, 79, 83, 93, 131–3, 155, 191–3, 231; capture and storage (CCS) 80–1, 83; price 79, 154–5; politics of 82–4; black 188–93
carbon dioxide emissions 5, 10–12, 20, 21, 52–84, 93–5, 100–1, 105, 116, 122, 123–4, 127, 128, 130, 131–2, 133, 135, 149–50, 154–5, 167, 180, 184, 188, 191, 198–9, 201, 203, 204, 205, 206, 208–9, 211, 212, 215, 227, 236, 238, 242, 243: planetary boundary for 52–84
Carbon Trade Watch 192
cars *see* vehicles
Cartagena Dialogue 242
Cato Institute 210
CFCs 217–28
Cheatneutral.com 192
Chernobyl 170–6, 177
Chesser, Robert 175
China 21 coal power in 11, 80, 149–50, 178–9, 185, 190, 239; nuclear power in 71, 243; dam construction 71, 140; 'nightsoil' industry 99; meat eating in 120; demand for fossil fuels 123; alternatives to high carbon aviation 124; hydroelectricity 150; virtual water and 153; pollution incidents 166; aerosol pollution 185, 187; black carbon and 189; transport pollution 190; emissions standards 190; CFC production 224; Copenhagen summit and 231, 232, 233, 234; population growth 236, 237; vehicle ownership, growth in 238; emissions 239–40; food production 240; investment in low-carbon technologies 242–3
Chinese Academy for Environmental Planning 185
Climate Action Network 133
Climate Action Partnership 84
climate change: carbon offsetting/markets and 1, 11, 50, 79, 83, 93, 131–3, 155, 191–3, 231; carbon dioxide emissions and *see* carbon dioxide emissions; deniers 6, 7–8, 16, 172, 210–16, 226–7; extinction and 33; boundary *see* climate change boundary; tipping points 60–4; methane and 17, 63–4, 94, 101, 155, 204, 205; population growth and *see* population; nuclear power and *see* nuclear power; renewables and *see* renewables; agreements/negotiations 82–3, 133, 226, 225–34, 242, 243 *see also under individual agreement/negotiation name*; nitrates worsen 94; solar radiation management and 194–7; China and *see* China *see also under individual event and area name*
climate change boundary 12, 52–84, 96, 125, 128, 149, 150, 180, 198, 200, 208, 209, 235; 350: current evidence 54–9; 350: modelling evidence 60–4; 350: past evidence 64–6; towards a technofix? 66–9; technologies for 350 69–78; new technologies for the future 78–81; politics of carbon 82–4; sea level rise 62; Arctic thaw and 56–8, 60–2; destabilisation of Atlantic Ocean circulation 62; models 64–5
Climate Fix, The (Pielke Jnr) 68
'Climategate', 2009 226–7
Climatic Research Unit, University of East Anglia 227
Clinton, Bill 225
Club of Rome 9, 237
coal power 10, 11, 21, 52, 66, 70, 71, 72, 73, 74, 80–1, 84, 96, 125, 129, 130, 132, 149–50, 167, 178–9, 180, 181, 184–5, 187, 188, 190, 191, 227, 234, 237, 239
Cochabamba, Bolivia 153–4
Collapse (Diamond) 111
Colorado River 57, 58, 139, 145, 148
Commission for the Conservation of Atlantic Tunas 40
Condit Dam 147
Congo Basin Forest Fund 131
Congress, US 53, 84, 223
Convention on Biological Diversity, Nagoya, 2010 50–1, 115–16, 117
Convention on Long-range Transboundary Air Pollution, 1979 96
COP15 of the UN Framework Convention on Climate Change 229
Copenhagen Accord 232, 233
Copenhagen climate summit, 2009 82–3, 226, 229–34, 242, 243
coral reefs 12, 18, 22, 31, 40, 44, 45, 46, 47, 49, 59, 199, 200–1, 202, 203, 204, 205–6, 207, 208, 209
Corporate Watch 106
Costa Rica 69, 136, 163, 242
Costanza, Robert 47
Cretaceous Period 24, 25–6, 37, 65, 204, 205
Crookes, William 87–8, 108
Crutzen, Paul 194–6, 219
Current Opinion in Plant Biology 107
Cyclone Nargis 146

Da Silva, Luiz Inácio Lula 116
Dai, Aiguo 195
Daly, Herman 240
Dampier, William 39

dams, removing unnecessary 12, 147–8; hydroelectric 45, 73, 77, 130, 144, 148, 149–50, 239; Chinese construction of 71, 140; fishery collapse and 94; tidal barrages 127; block natural flow of water 139, 140, 142, 144; threaten species 143–5; affects water temperature 145; water trapped behind loses most of it's sediment load 145–6; current water use 146; where water is taken from 146–7
Danish Committee on Scientific Dishonesty 176
DDT 158, 159, 160, 162–3, 165, 218
Dead Sea 144, 150
dead zones 21, 92, 93, 97, 202, 218
deep-sea floating turbines 129–30
deforestation 49, 58, 63, 111, 112, 116–17, 122, 130–4, 137, 140
Delta smelt 143
'demographic transition' 135–8
Dhaka 145–6
Diamond, Jared 111
diesel engines 69–70, 76, 96, 137, 187, 188, 189
Dinorwig, Wales 150
DuPont 221, 222, 227

earth: goldilocks state 15–16; self-regulating 16–17, 18–20; carbon cycle *see* carbon: cycle; 'snowball' 17; ice-albedo feedback 18
'earthshine' 194
East Antarctic Ice Sheet 62
Economics of Ecosystems & Biodiversity, The (*TEEB*) report, 2010 48
Economist 77
Ecuador 131, 136
Edwards Dam 147
Egypt 73, 152
electric vehicles 80, 122–3, 124, 190
Endangered Species Act, US 57
Energy Information Administration, US 74
'energy sprawl' 129, 130
'enigmatic decline' 163–4
Enriching the Earth (Smil) 88
Environmental Protection Agency, US 163, 223–4
Eocene 12, 65, 205
ethanol 121–2
European Emissions Trading Scheme 79
European Union 53, 77, 156, 165, 166, 189, 190, 225
evolution: intelligence/brain 5, 6, 24, 25, 26, 27–8; as part of process of Earth's self-regulation 18–20; of life 23–6; adaptation to a changed environment and 206
extinction: mass 6, 18, 19, 20, 25–6, 30, 37, 45–6, 137, 204–8; animal 31–45, 51, 57, 127, 137, 143, 158–9, 163, 164
Eyjafjallajokull volcano eruption, 2010 7

farming *see* agriculture
Federal Endangered Species List, US 144
Feely, Richard 198
fertiliser, agricultural 88–109, 141–2, 240
fire-ape, birth of 27–9
'Flask world' 20, 21
'flood basalts' 204
Food Animal Initiative 119–20
food-fuel relationship 29
France: nuclear power in 71, 75, 180; wheat yields in 155, 156; Copenhagen summit 232; emissions 71, 239
freshwater boundary 139–56; fragmentation of rivers by man-made infrastructure 139–40; benefits of water engineering and control 140–2; ecological zones 142–7; dams and 140, 144–50; removal of water from rivers for agriculture 148–9; hydroelectric dams 144–5, 149–50; desalination 150–1; extinction rate of freshwater species 143–4; water temperature 145; storm surges 146; freshwater ecological restoration 147–8; increasing global trade in virtual water 152–3; privatisation of 153–5; water wastage 154, 155; putting a price on water 154–5
Friends of the Earth 83, 102, 127, 132, 179, 181
Fuji-Hakone-Izu Park, Japan 111
Fukushima disaster, 2011 10–11, 71–2, 167, 170, 176–8, 179, 181, 182

G8 232
Gaia theory 16–18, 20, 218
Galapagos Islands 38, 39, 41
Galiana, Isabel 79
Ganges River 146
'gap species' 117
GDP as measure of progress 67–8
'gender-benders' 159, 163–4
genetic engineering (GE) 3–4, 11, 12, 101–9, 119, 214
geoengineering 11–12, 194–7, 219
Germany 10–11, 56, 65, 70, 71, 72, 77, 90–1, 111, 128, 170, 182, 191, 195
'ghost habitats' 36
Gingrich, Newt 7
glaciers 59, 151, 187, 188, 227
Glen Canyon Dam 145
Glines Canyon Dam 147
Global Biodiversity Outlook 31, 46
'global dimming' 186, 193
Global Warming Policy Foundation 210, 212
Global Wind Energy Council 74
glyphosate 104, 105
gomphotheres 36
Gooding, Mike 120
Google 84; Foundation 74
Gore, Al 225
Gorham's Cave, Gibraltar 34
Gothenburg Protocol, 1999 96
GRACE (Gravity Recovery and Climate Experiment) 61–2
Grand Teton National Park, Wyoming 44
great apes, extinction of 32–3
Great Barrier Reef, Australia 200–1
Green Fund 78
Green Investment Bank 77
Green Revolution 92, 98, 141–2, 155, 156
Green, Chris 79
Greenland 56, 60–2, 65, 66
Greenpeace 102, 122, 127, 158, 166, 167, 172–3, 181, 211
Greens 6, 10, 11; biodiversity loss and 46; call into question GDP as measure of progress 67–9; push the difficulty of dealing with climate change 82, 214–15; genetic

engineering and 102–3; plea to give up flying 124; campaign against environmental toxins 161; nuclear power and 180–2; anti-science 214; population growth and 237
Guardian 42, 102–3, 108, 181
Gulf of Mexico 21, 92, 97

Haber, Fritz 88–91
Haber-Bosch process 88–92, 94, 109, 217–18
HadCM3 63
Hambler, Clive 127
Hansen, James 53, 57, 66, 181
Helm, Dieter 239
Himalayas 32, 59, 151, 185, 187, 227
History of Life, The (Cowen) 34
Ho Chi Minh City 146
Holocene 5, 12
hominids 22, 26–8, 34, 36
Homo erectus 27, 34
Homo habilis 27
Homo neanderthalensis 27, 34
Homo pyrophilus 29
Homo sapiens: dominance of 5–6; ascent of 15–22; god species or rebel organism? 18–22; descent of 22–6; birth of 27–9; brain development 27–8; drive hominid relatives to oblivion 33–4; Pleistocene overkill and 33–4
Hudson River 155
Huhne, Chris 128
Hurricane Katrina 146
Hybrid: the history and science of plant breeding (Kingsbury) 106
hydroelectric dams/power 45, 77, 130, 144–5, 148, 149–50, 239
hydrogen as fuel 80
hydrological cycle 142, 186–7
hydrological engineering 139, 141

Independent 127, 181
India: carbon emissions 11; alternatives to high carbon aviation 124; hydroelectricity in 150; vultures in 158–9; pollution in 185; black carbon and 189, 191; Copenhagen summit and 234
Indonesia 115, 122, 131
Intergovernmental Science-Policy Platform on Biodiversity and Ecosystem Services (IPBES) 51
International Energy Agency (IEA) 78
International Institute of Tropical Agriculture 107
International Monetary Fund (IMF) 78
International Whaling Commission (IWC) 38
IPCC (Intergovernmental Panel on Climate Change) 51, 58, 62, 64, 227
irrigation 119, 140, 141–2, 148, 152
Israel 123, 144, 150, 155, 163
IUCN Red List 32, 37–8, 143, 160

J. Craig Venter Center 3
Jackson, Jeremy 203, 240
Jacobson, Mark 188
Japan: earthquake and tsunami, 2011 7; climate change in 59; Fukushima disaster, 2011 10–11, 71–2, 167, 170, 176–8, 179, 181, 182
Jordan River 144, 150, 155
Journal of Geophysical Research 57

keystone/apex predators 31, 43, 44
Kruger, Tim 209
Kunin, Professor Bill 100
Kyoto Protocol, 1997 133, 225–8, 236, 243

Labrador Sea 56
Lake District 56, 110, 111, 171
Lake Powell 145
land use boundary 110–38; human impact on 110–12; importance of ecological zones 113–15; a plan for land 115–20; protected land 115–20; global value of wilderness 118; intensification of farming and 118–20; integrated pest management 119; agro-forestry 119; aquaculture 119; meat and energy 120–30; REDD 130–4; population growth and urbanisation 134–8
Leipold, Gerd 172, 173
Lenton, Tim 19, 20, 60, 62
Les Rois 34
Liberal Democratic Party 169
life, creating new forms of 3–5; origin of 23–6
Limits to Growth report, 1972 9, 238
Lovelock, James 16, 20, 218–19
Lovins, Amory 76, 125
Lucas, Caroline 214–15

Madagascar 36, 59
Mahli, Yadvinder 63
malaria 141, 158
Maldives: levy on diving trips considered 49; climate change and 61, 69–70, 76; Copenhagen summit and 230, 232, 233; pledges carbon neutrality 242
Malua Forest Reserve 49–50
mass extinctions 6, 18, 19, 25–6, 30, 37, 45–6, 137, 204–8
Max Planck Institute 195
McKibben, Bill 53
McNeill, John 217, 218
megacities 123, 134, 145–6
'megafauna fruit' 36–7
Mekong delta 146
mercury 158, 160, 167, 180
Merkel, Angela 232
Met Office Hadley Centre, UK 63
methane 17, 63–4, 94, 101, 155, 204, 205
Mexico 25, 40, 41, 49, 136, 145, 146, 163, 230
Michaels, Patrick 210
Midgley, Thomas 217, 218, 219
Mississippi river 139
Missouri River 144
'mitigation hierachy' 50
Mitsubishi 40
Molina, Mario 219
Möller, Anders 176
Moloney, Maurice 107
Monbiot, George 72, 181
Monsanto 104, 106
monsoon 140, 185, 186–7, 195
Montreal Protocol, 1987 220–4, 225, 228, 243
Mousseau, Tim 176
mudflats 127
Muller-Landau, Helene 137

NASA 53, 55, 57, 66, 181, 219, 220, 222
Nasheed, President 69, 230, 232, 233
national parks 58, 110, 111, 112, 131, 160

National Science Academy, US 107
National Snow and Ice Data Center (NSIDC) 56
National Trust 127
National Water Carrier, Israel 144
natural capital 30–1, 48, 133–4
Nature 219, 235
Nature Conservancy 129, 1148
New Economics Foundation (NEF) 68–9, 214–15, 238
New Forests 49
New Internationalist 192
NGOs 11, 153, 167, 181
Niger, River 146
Nile, River 145
nitrogen boundary 85–109: famine and 86, 87–8; Haber-Bosch process 88–91, 94, 109; nitrogen ape 91–5; meeting the 95–100; human production of nitrogen, benefits of 93–4; nitrogen oxide emissions 94–6; efforts to control nitrogen pollution 96–100; microbial denitrification 97; wastewater, removing nitrates from 97–8; organic farming and fertiliser use 98–101; designing crops that are more efficient in nitrogen uptake 101–9
North America: extinction in 34–5, 43; 'rewilding' 36; climate change in 57–8; GE in 104; land use in 117, 119; wind power in 126
Northern permafrost zone carbon store 63–4
Norway 38, 131
no-till agriculture 105, 119
nuclear power 10–11, 12, 70–3, 75, 79, 81, 124–5, 127–8, 129, 130, 209, 214, 215, 243; pollution/dangers of 167–82
Nuon Renewables 126–7

Obama, Barack 7, 232, 243
ocean acidification 8, 113; waste in 21; evolution of 24; animals depleted 37–41; boundary 198–216; life in acidic oceans 200–3; reef gaps 203–6; oceans of the future 207–10; carbon cycle and 198–9; ocean pH 199–200; sea creatures and 202; changes characteristics of seawater 202–3; fertilising effect of 203; 'calcification crisis' 204; ground-up olivine rock spread as beach nourishment 208–9; addition of lime directly into oceans 209; attention to 210; sceptics 210–16
Open Atmospheric Science Journal 66
Overpeck, Jonathan 57, 58
Oxford University 1, 63, 77, 84, 127, 239
ozone layer 12, 93, 95, 185; boundary 54, 217–28

Pachauri, Rajendra 227
Pakistan 55, 140, 153, 179
Palaeocene-Eocene Thermal Maximum (PETM) 205
palm-oil companies 46, 122, 131
Papua New Guinea 42, 163
Parliamentary Environmental Audit Committee, UK 185–6
PCBs 12, 159, 160, 163, 165
Peacock, Jim 107
peak oil 237
Permian Period 204, 205
pesticides 104–7, 118, 119, 141–2, 158, 163, 165, 218
Petrovich, Leonid 173, 175
photosynthesis 18, 23, 202, 203
Pielke Jnr, Roger 68
Pinatubo eruption, 1991 185, 195
planetary boundaries, concept of 8–10, 12, 234–44; expert group 9, 19, 30, 47, 52, 53, 57, 60, 62, 66, 95, 115, 118, 119, 146, 164, 165, 184, 195, 198, 208, 219, 220, 234, 235; summary of 235–6; what about population? 236–7; do not deal with resource constraints 237–41
plastics 4, 5, 154, 157, 166, 241
Pliocene epoch 65
polar bear 19, 22, 57
pollution *see under individual pollutant name*
population, human 6, 21, 22, 31, 67, 69, 86, 87, 92, 111, 112, 115, 120, 134–8, 140–2, 146, 151, 236–7, 238, 242
Porritt, Jonathan 127, 236
Pretty, Jules 119
Prosperity Without Growth (Jackson) 240
pumped storage reservoirs 80, 150

Quaternary Megafaunal Extinction 37, 42

rainforest 11, 12, 30, 42, 46, 62–3, 100, 116–17, 122, 131, 134, 187
Rasmussen, Lars Lokke 230–1, 232
Rational Optimist, The (Ridley) 210, 212
'RE‹C' 84
REACH (Registration, Evaluation, Authorisation and Restriction of Chemicals), 2007 165
Reagan, Ronald 223
RealClimate blog 64
'rebel organisms' 20–2
Reducing Emissions from Deforestation and Degradation (REDD) 130–4
renewables 70, 72–81, 83, 84, 124–30, 142, 149, 150, 151, 155, 178, 180, 181, 242–3
'rewilding' 36, 133, 136
Ricardo, David 153
Ridley, Matt 113, 210, 211, 212
Rio Grande 145
rivers: dams and reservoirs on 12, 130, 139–40, 144–6 *see also under individual dam or reservoir name*; drying of 63; eutrophication 92; nitrogen in 94, 97, 101; ecological zones 142–3; agricultural use of water from 148–51; 'gender-benders' in 159; toxics in 159, 161, 166
Roberts, Callum 39
Rockström, Professor Johan 8, 9
Rowland, F. Sherwood (Sherry) 219
Royal Academy of Engineering 71
Royal Society 207
Rozzi, Fernando 34
RSPB 127
Rudd, Kevin 232
runoff 92, 94, 95, 105, 120, 140, 151, 195, 235

Sahara 73, 128
Sahel, Africa 186
Sarkozy, Nicolas 232
Saudi Arabia 150–1, 152, 232
Schellnhuber, Hans Joachim 60

Science 57
sea: depletion of animals in 37–41; levels 35, 61–2, 65, 69, 139 *see also* ocean acidification
Serreze, Mark 56
Severn Estuary tidal barrage 127, 128
ship-emitted pollutants 189–90
Siberia 36, 42, 56, 63, 64, 118, 205
Sierra Club 181
Silent Spring (Carson) 158
Sites of Special Scientific Interest 120
Six Degrees (Lynas) 53, 197
Sky, Jasper 77, 78
Smil, Vaclav 88, 99
Snyder, Peter 114
solar power 70, 73–6, 77, 79, 125, 128–9, 130, 151, 166–7, 178, 183, 191, 209, 227
solar radiation 70, 183, 185, 194–7
soot particle 183, 185, 186, 187, 188–93
'species-area relationship' 45
Stanford University 100, 188
State of the Air 2010 report (American Lung Association) 186
Stern Review on the Economics of Climate Change, 2006 53
Stockholm Convention on Persistent Organic Pollutants, 2001 165
Stockholm Resilience Centre 8
stoves, dirty cooking 191, 193
Sukhdev, Pavan 48
sulphur emissions 96, 183, 184, 185, 186, 188, 190, 193–7
Sumatra 122
'supergrid' 77, 130
Sustainable Development Commission 127
'sustainable development' schemes 136
Sustainable Rivers Project 148
Suzlon 74
synthetic biology 3–4, 11, 12
Syr Darya 149

'Target Atmospheric CO2: Where Should Humanity Aim?' 66
Tea Party 212, 213
'terra preta' 111
Thames, River 120, 140, 159, 166
The New Home Front report, 2011 214–15
Three Gorges Dam, China 149
Three Mile Island 177
tiger 32, 42, 43, 51
Tindale, Stephen 181
toxics 8,12, 157–82; pesticides 104–7, 118, 119, 141–2, 158, 163, 165, 218; plastics 157; mercury 158, 160, 167; DDT 158, 159, 160, 162–3, 165; Indian vulture and 158–9; 'gender-benders' 159, 163–4; accumulate in food chain 159–60; accumulate in polar regions 160; human risk 160, 161–3; 'enigmatic decline' 163–4; boundary/regulation 164–7; nuclear pollution 167–82; PCBs 159, 160, 163, 165; atrazine 159–60, 161, 163–4
transgenic crops 102, 104, 107, 108, 119
Trenberth, Kevin 195
Triassic period 24, 204, 206
'trophic cascade' 43–4

Udall, Bradley 57, 58
United Arab Emirates 152
United Kingdom: wind power in 74–5, 77, 130; Climate Change Act 83; restored wildflower meadows 94; organic farming in 100; genetically engineered crops in 103; land use in 110; Severn Estuary tidal barrage 127; VAT rise in 133–4; radon in 168; CFCs in 223; carbon dioxide emissions 240
United Nations 135, 137; Food and Agriculture Organisation 50; Framework Convention on Climate Change 50–1, 53, 229, 231; Millennium Development Goals 98, 141; Population Fund 134; Scientific Committee on the Effects of Atomic Radiation (UNSCEAR) 171, 172
United States: solar power 73, 74, 128–9; agricultural fertiliser, dependence on 92; meat consumption 120–1; ethanol production 121–2; freshwater species extinction 143; freshwater ecological restoration 147–8; seasonality of runoff 151; nuclear power in 180; climate change deniers in 212; CFC regulation and 223–4; Kyoto Protocol and 225; Copenhagen summit, role in 233–4; clean energy and 243
urbanisation 134–8
US Army Corps of Engineers 144, 148
Uzbekistan 148–9

valueing natural assets 46–50
VAT, adding half a per cent to 133–4
vehicles: efficiency of 76: electric 80, 96, 122–3, 124, 190; diesel 96, 189; rate of ownership 238
Venter, J. Craig 3, 4, 5, 12, 23
Vestas 74
Vienna Convention for the Protection of the Ozone Layer, 1985 220, 222–3
Vietnam 136–7
Volga, river 145

Watt, James 21
Wen Jiabao 231
West Antarctic Ice Sheet 61
wetlands 49, 97, 111, 113, 127, 133, 142, 143, 148, 243
whaling industry 37–8
Whole Earth Discipline (Brand) 103–4
wildfires 58, 187, 189
Williams, Hywell 19, 20
Williams, Steve 58–9
Wilson, E. O. 43
wind power 70, 71, 74–5, 76, 77, 79, 85, 124–7, 130, 150, 151, 178, 242–3
World Bank 50
World Future Council 78
World Health Organisation (WHO) 97–8, 141, 185, 191
World Meteorological Organisation 222
World Trade Organisation (WTO) 153
Wright, Joseph 137
WWF (World Wildlife Fund) 158, 193

Yakama Indian Nation 147
Yangtze River 139, 140, 145, 149, 150
Yasuni National Park 131
Yellow River, China 144
Yellowstone National Park 44, 111–12
Yu Qingtai 231, 232, 233